Christoph Kolodziejski

Mathematical Description of Differential Hebbian Plasticity

Christoph Kolodziejski

Mathematical Description of Differential Hebbian Plasticity

and its Relation to Reinforcement Learning

Südwestdeutscher Verlag für Hochschulschriften

Impressum/Imprint (nur für Deutschland/ only for Germany)

Bibliografische Information der Deutschen Nationalbibliothek: Die Deutsche Nationalbibliothek verzeichnet diese Publikation in der Deutschen Nationalbibliografie; detaillierte bibliografische Daten sind im Internet über http://dnb.d-nb.de abrufbar.

Alle in diesem Buch genannten Marken und Produktnamen unterliegen warenzeichen-, marken- oder patentrechtlichem Schutz bzw. sind Warenzeichen oder eingetragene Warenzeichen der jeweiligen Inhaber. Die Wiedergabe von Marken, Produktnamen, Gebrauchsnamen, Handelsnamen, Warenbezeichnungen u.s.w. in diesem Werk berechtigt auch ohne besondere Kennzeichnung nicht zu der Annahme, dass solche Namen im Sinne der Warenzeichen- und Markenschutzgesetzgebung als frei zu betrachten wären und daher von jedermann benutzt werden dürften.

Verlag: Südwestdeutscher Verlag für Hochschulschriften Aktiengesellschaft & Co. KG
Dudweiler Landstr. 99, 66123 Saarbrücken, Deutschland
Telefon +49 681 37 20 271-1, Telefax +49 681 37 20 271-0
Email: info@svh-verlag.de
Zugl.: Göttingen, Georg-August-Universität Göttingen, Dissertation, 2009

Herstellung in Deutschland:
Schaltungsdienst Lange o.H.G., Berlin
Books on Demand GmbH, Norderstedt
Reha GmbH, Saarbrücken
Amazon Distribution GmbH, Leipzig
ISBN: 978-3-8381-1372-2

Imprint (only for USA, GB)

Bibliographic information published by the Deutsche Nationalbibliothek: The Deutsche Nationalbibliothek lists this publication in the Deutsche Nationalbibliografie; detailed bibliographic data are available in the Internet at http://dnb.d-nb.de.

Any brand names and product names mentioned in this book are subject to trademark, brand or patent protection and are trademarks or registered trademarks of their respective holders. The use of brand names, product names, common names, trade names, product descriptions etc. even without a particular marking in this works is in no way to be construed to mean that such names may be regarded as unrestricted in respect of trademark and brand protection legislation and could thus be used by anyone.

Publisher: Südwestdeutscher Verlag für Hochschulschriften Aktiengesellschaft & Co. KG
Dudweiler Landstr. 99, 66123 Saarbrücken, Germany
Phone +49 681 37 20 271-1, Fax +49 681 37 20 271-0
Email: info@svh-verlag.de

Printed in the U.S.A.
Printed in the U.K. by (see last page)
ISBN: 978-3-8381-1372-2

Copyright © 2010 by the author and Südwestdeutscher Verlag für Hochschulschriften Aktiengesellschaft & Co. KG and licensors
All rights reserved. Saarbrücken 2010

*Dedicated to my mother Lydia,
and my grandmother Gertrud*

Abstract

The human brain consists of more than a billion nerve cells, the neurons, each having several thousand connections, the synapses. These connections are not fixed but change all the time. In order to describe synaptic plasticity, different mathematical rules have been proposed most of which follow Hebb's postulate. Donald Hebb suggested in 1949 that synapses only change if presynaptic activity, i.e. the activity of a synapse that converges to the neuron, and post-synaptic activity, i.e. activity of the neuron itself, correlate with each other. A general descriptive framework, however, is yet missing for this influential class of plasticity rules. In addition, the description of the dynamics of the synaptic connections under Hebbian plasticity is limited either to the plasticity of only one synapse or to simple, stationary activity patterns. In spite of this, Hebbian plasticity has been applied to different fields, for instance to classical conditioning. However, the extension to operant conditioning and to the closely related reinforcement learning is problematic. So far reinforcement learning can not be implemented directly at a neuron as the plasticity of converging synapses depends on information that needs to be computed by many neurons. In this thesis we describe the plasticity of a single plastic synapse by introducing a new theoretical framework for its analysis based on their auto- and cross-correlation terms. With this framework we are able to compare and draw conclusions about the stability of several different rules. This makes it also possible to specifically construct Hebbian plasticity rules for various systems. For instance, an additional plasticity modulating factor is sufficient to eliminate the auto-correlation contribution. Along these lines we also generalize two already existing models, a fact which leads to a novel so-called Variable Output Trace (VOT) plasticity rule that will be of further importance. In a next step we extend our analysis to many plastic synapses where we develop a complete analytical solution which characterizes the dynamics of synaptic connections even for non-stationary activity. This allows us to predict the synaptic development of symmetrical differential Hebbian plasticity. In the last part of this thesis, we present a general setup with which any Hebbian plasticity rule with a negative auto-correlation can be used to emulate temporal difference learning, a widely used reinforcement learning algorithm. Specifically we use differential Hebbian plasticity with a modulating factor and the VOT plasticity rule developed in the first part to prove their asymptotic equivalence to temporal difference learning and additionally investigate the practicability of these realizations. With the results developed in this thesis, it is possible to relate different Hebbian rules and their properties to each other. It is also possible for the first time to calculate plasticity analytically for many synapses with continuously changing activity. This is of relevance for all behaving systems (machines, animals) whose interaction with their environment leads to widely varying neural activation.

*"Wisdom lies neither in fixity nor in change,
but in the dialectic between the two."*

Octavio Paz (1914 - 1998)

Contents

Abstract		i
1 Introduction		**1**
1.1	Plasticity mechanisms and their relation to learning	1
1.2	Open loop versus closed loop	6
1.3	Neuronal activity	8
1.4	Definitions and roadmap	10
2 Single-Plastic-Synapse Systems		**13**
2.1	S&B model, ISO learning and VOT plasticity	15
2.2	Hebb learning	22
2.3	TD learning	23
2.4	ICO learning	27
2.5	ISO3 learning	28
2.6	Discussion	31
3 Many-Plastic-Synapse Systems		**35**
3.1	Multiple plastic synapses for a *single* input	35
3.2	Multiple plastic synapses for *many* inputs	40
	3.2.1 Symmetrical rules: ICO learning	40
	3.2.2 General many-synapse systems	42
	3.2.3 Symmetrical rules: ISO learning	47
3.3	Discussion	49
4 Relation to Reinforcement Learning		**51**
4.1	General setup	54
4.2	General analysis	57
	4.2.1 Global third factor	63
	4.2.2 Local third factor	69
	4.2.3 Different time scales: VOT plasticity	75
4.3	Discussion	80
5 Discussion and Outlook		**85**

A	**Biophysical Basics**	**97**
B	**Quasi-Static Weight Changes**	**99**
	B.1 General solution	99
	B.2 Quasi-static solution	100
	B.3 Variation of parameters	100
C	**Numerical Considerations**	**103**
D	**Solution of the Homogeneous Part of the General Differential Hebbian Plasticity Equation**	**105**
E	**Switching Integral and Derivative to Solve the Derivative of the Exponential Integral**	**107**
F	**Estimation of the Number of Calculations for Numerical Calculation of the Temporal Development**	**109**
G	**Solution of the Difference Equation Given by the Overall Weight Development**	**111**
H	**Analytical Calculation of γ Using First and Second Order Terms**	**115**
	H.1 Taylor expansion of the kernel function	115
	H.2 Intervals given a third factor	116
	H.3 Analytical calculation of κ_G and κ_L	117
	H.4 Analytical calculation of $\tau_G^\pm$ and τ_L	119
	H.5 Analytical calculation of γ_G and γ_L	125
	H.6 Analytical calculation of κ_T, $\tau_T^\pm$ and γ_T	125
	H.7 Analytical calculation of γ_T for the S&B model	127
	List of Symbols	**129**
	Bibliography	**131**
	Acknowledgments	**141**

Chapter 1

Introduction

In this thesis we are mainly concerned with the mathematical details of differential Hebbian plasticity and its relation to learning. This chapter will introduce these aspects asking first of all what the difference is between them? Plasticity means that there is a deformable or shapeable entity and in the nervous system this refers to the connections between the neurons which can change. Connections can totally relocate (structural plasticity - Chklovskii et al. (2004); Fox and Wong (2005); Butz et al. (2008)) but in the context of this work, plasticity stands for the variation of the connection *strength*. Learning on the other hand is a more abstract term. It is used in a general way in school, e.g. when you memorize vocabulary or when trying new sports in order to acquire new skills. Thus, learning is a word used at the level of behavior. In the following we will start with plasticity and afterwards try to link these rather biological mechanisms to behavior.

1.1 Plasticity mechanisms and their relation to learning

In 1949 Donald Hebb (Hebb, 1949) wrote a statement that is still influencing the neurosciences:

> When an axon of cell A is near enough to excite cell B and repeatedly or persistently takes part in firing it, some growth process or metabolic change takes place in one or both cells such that A's efficiency, as one of the cells firing B, is increased.

A simple equation can be deducted from Hebb's idea: $\Delta w_{BA} = u_A \cdot u_B$, where w stands for the efficiency of a connection between cell A and B, and u_A and u_B for the activity of A and B respectively. There would be no changes induced in the efficiency of the connections between cell A and B whenever only one of the two cells is active. Only if both cells are active at the same time, thus when both activities coincide, efficiency increases. That is also the reason, why these kind of plasticity rules are called correlation rules or coincidence detecting rules.

In the following we exchange the old terminology of "cells" and "connections" by the modern "neurons" and "synapses". The efficiency of a connection is therefore the synaptic strength, or in short *weight*. The name weight results from the fact that a neuron usually has more than one

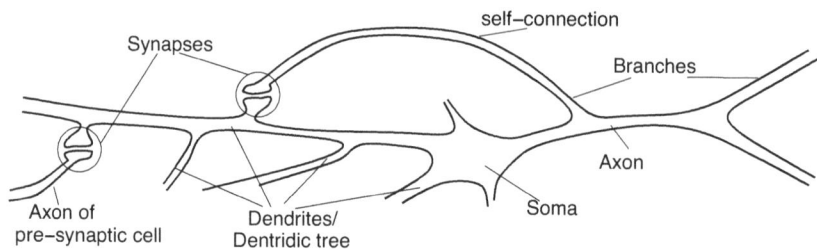

Figure 1.1: *Basic structure of a neuron (which has a self-connection). Usually information arrives at the dendritic tree and is summed up at the soma. From there on information reaches other neurons transferring a signal via the axon (which can also split into branches). The connections between axons and dendrites are called synapses.*

synaptic connection and the synaptic efficiency tells you how to weigh the activity of neurons connected to a specific target neuron. This idea of a neuronal network is depicted in more detail in Figure 1.2 left. A neuron consists not only of the cell body, the soma, as Figure 1.2 left may suggest, but also of parts that collect activity from other neurons, the dendrites, which belong to the dendritic tree and it has an axon, which transmits the activation to other neurons or even back to itself (recurrent connection). At the soma the activity of all dendrites is summed and further processed. This is sketched in Figure 1.1.

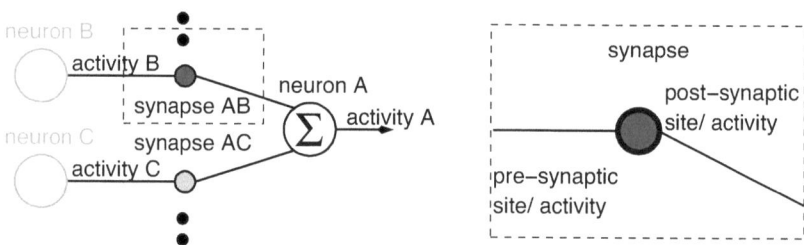

Figure 1.2: *Basic scheme of a neuronal network. On the left a single neuron A receives activity from different neurons (among others B, C) via synapses (among others AB, AC). The right panel shows a close up of one synapse. The site where activity reaches the synapse is called pre-synaptic site and that where activity is collected by the neuron is called the post-synaptic site.*

Hebb made his postulate in the middle of the last century, and it has long remained unknown whether such a mechanism would really exist in the nerve cells of our brain. In 1973 Bliss and Lømo were the first to report a mechanism called long-term potentiation (LTP) which is directly related to Hebbian plasticity. The first problem Bliss and Lømo faced was how to measure the synaptic efficiency. It turns out that the only way to do this is by measuring an excitatory post-

1.1 PLASTICITY MECHANISMS AND THEIR RELATION TO LEARNING

synaptic potential (EPSP) which is the positive activity at the post-synaptic site of a synapse (see right panel of Figure 1.2 for details and Figure 1.3 B for example EPSPs). Bliss and Lømo measured an EPSP after they activated the pre-synaptic site and compared this control result to the same measurement, however now using high-frequency ($\sim 100\,Hz$) stimulation of the pre-synaptic site (compare with second stimulus of Figure 1.3 A). The EPSP amplitude increased after the high-frequency stimulation. Bliss and Lømo had only stimulated the pre-synaptic site (cell B in our introductory example). Still, the found effect can be related to Hebbian plasticity, because high-frequency activation at the pre-synaptic site drives activity also at the post-synaptic site (hence at cell A in our introductory example) leading to the required pre-post correlation.

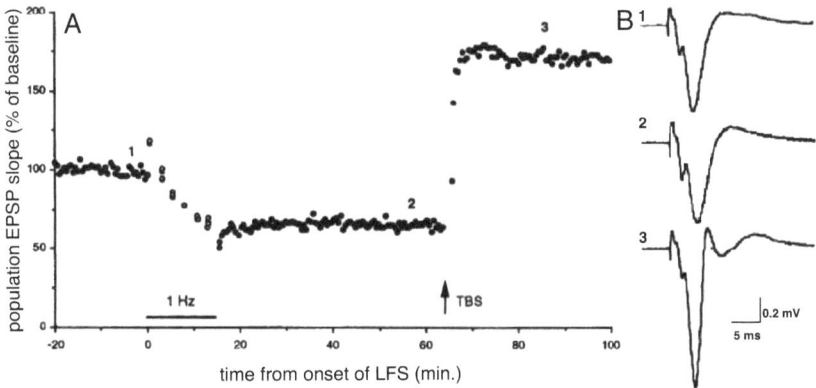

Figure 1.3: *Measurements of LTD and LTP in the Schaffer collateral-CA1 pathway in the Hippocampus. In panel A amplitudes of the EPSP are plotted against time. At 0 minute a low frequency ($\sim 1\,Hz$) stimulus was given to the collaterals, and the EPSP amplitude decreased. After about one hour a high frequency stimulus (TBS - theta burst stimulus - $\sim 100\,Hz$) was applied which leads to an increase. In panel B particular EPSPs (negative peaks) are shown for the indicated times. Recompiled from Dudek and Bear (1993).*

After the discovery of LTP, many theoreticians suggested that also a decrease in efficiency should take place at synapses, long before Dudek and Bear in 1992 finally found a reduction in synaptic efficiency, called long-term depression (LTD). To capture LTD by the basic Hebb rule, it was altered in many different ways in order to incorporate negative changes in the synaptic strength. The three most prominent ways to do this are *anti-Hebb* (e.g. Lisman (1989)) where just a minus sign is included, *threshold* (or *covariance*) *models* (e.g. BCM rule by Bienenstock et al. (1982)) where a threshold is introduced. Here either pre- or post-synaptic activity needs to exceed the threshold in order to drive positive weight changes whereas otherwise, changes are negative as intended. The last way by which weights decrease is achieved by a decay (or leakage) term (e.g. Oja (1982)) that drives the synaptic efficiency to zero without any activity.

This method also leads to so-called weight-normalization for which indirect evidence was also found later (Bi and Poo, 1998).

How did Dudek and Bear (Dudek and Bear, 1992) achieve a negative change in the efficiency? They varied the frequency of the stimulations at the pre-synaptic site and found that long-lasting low frequency ($\sim 1\,\text{Hz}$) stimuli induced negative changes (see the first stimulus of Figure 1.3 B). To verify that the cause for the change of synaptic strength was also at the post-synaptic site, Feldman altered the post-synaptic potential by a special technique (voltage patch clamp) while stimulating the pre-synaptic site. Without a change at the post-synaptic site the EPSP amplitude remained constant. However, by putting the potential to different levels he either increased or decreased the amplitude of the EPSP (Feldman, 2000).

The level that we have covered until now was only phenomenological, describing methods responsible for changes in the synaptic strength. What really happens at, or rather within a synapse is, however, not so clear, yet[1].

Most often responsible for synaptic plasticity at the post-synaptic site are so-called N-Methyl-D-Aspartat ion-channels (NMDA, Malenka and Nicoll (1999); Dudek and Bear (1992)) and the Ca^{2+} ion (Yang et al., 1999; Bi, 2002). The NMDA channel is permeable to Ca^{2+} but only if a certain type of neurotransmitter (Glutamate) binds to it and a certain post-synaptic voltage level is reached. The first requirement is fulfilled whenever an action potential reaches the pre-synaptic site (see Figure 1.4 A). The latter holds if either the neuron at the post-synaptic site also produces an action potential or a total sum of the post-synaptic potentials of nearby synapses are high enough to produce a dendritic spike (Colbert (2001); Golding et al. (2002); see Figure 1.4 B). The Ca^{2+} ion depending on its concentration or rather the change in concentration (Yang et al., 1999; Bi, 2002) within the post-synaptic membrane then initiates a biochemical cascade which increases the number of NMDA channels. Note that this very short introduction has oversimplified the physiological complexity. The actual kind of plasticity (LTP or LTD) also depends on the synapse type, modulatory substances, the type of neurotransmitter used and the order in which the pre- and post-synaptic action potentials arrive at the synapse. Some of these aspects are of relevance for this study, most others will not be further considered.

An influence of the temporal signal order onto plasticity was proposed by Gerstner et al. (1996) and experimentally confirmed by Markram et al. (1997). Markram and his colleges found that not only the activity as such matters but also the timing. Whenever there is a spike at the post-synaptic site *after* there was a spike at the pre-synaptic site, the strength of the synapse increases (LTP). However, if the timing is acausal, which means there is a post-synaptic spike *before* there was a pre-synaptic spike, efficiency of the synapse decreases. This phenomenon is therefore called spike-timing-dependent plasticity (STDP). One way to model this aspect of plasticity is by including the change of post-synaptic activity: $\Delta w = u_B \cdot \Delta u_A$. If *pre* is before *post*, we correlate the pre-synaptic activity mostly with the rising phase of the post-synaptic activity and if *pre* is after *post*, we correlate pre-synaptic activity with the falling phase (see Figure 1.5 for a sketch or Figure 2.1 for a more detailed plasticity example). As changes at the post-synaptic site were used, this rule is called differential Hebbian plasticity.

---
[1] For details on the biophysics see appendix A

1.1 PLASTICITY MECHANISMS AND THEIR RELATION TO LEARNING 5

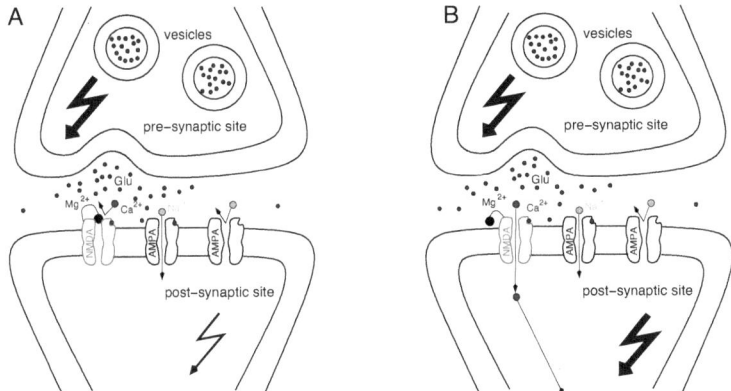

Figure 1.4: Sketch of a synapse with (A) and without (B) sufficient post-synaptic activity. Whenever there is pre-synaptic activity vesicles bind to the membrane and release their neurotransmitter (e.g. Glu - Glutamate). The transmitter binds to ion-channels which then become either open (AMPA[a]) or are still blocked (NMDA) by other ions (Mg^{2+}). As at least some of the ion-channels open, ions (e.g. Na^{2+}) can penetrate into the post-synaptic site, thus changing the potential (e.g. EPSP). However, in panel A the change in the potential is not enough to release the Mg^{2+} block at the NMDA channels which hinders Ca^{2+} to flow into the post-synaptic site. Only if there is a substantial depolarization (panel B) of the post-synaptic potential (e.g. because of a back-propagating action potential), the Mg^{2+} block at the NMDA is released and Ca2+ can flow into the post-synaptic region where it initiates different chemical reaction cascades that change the synaptic efficiency of this synapse.

---

[a]α-amino-3-hydroxyl-5-methyl-4-isoxazole-propionate

However, one important question remains. How would we know that plasticity, i.e. the change of synaptic efficiency, is really related to learning? In 2002 Martin and Morris in a review suggested four different criteria supporting that plasticity is the basic mechanism for learning: detectability, mimicry, anterograde and retrograde alternation (Martin and Morris, 2002). Detectability means that there are changes at the synapse levels after an animal has learned or memorized something. One example is the work of Rioult-Pedotti et al. (1998). In this study they prevented a rat from moving its left forelimb where the right one was freely movable while performing a skilled reaching task. Then they found that the EPSP amplitude of synapses in the motor cortex of the left hemisphere was higher compared to the right hemisphere. If we took the detailed results about the changes in the motor cortex of one rat and implemented these changes into another rat, then we would invoke the mechanism of mimicry. So far mimicry can not be experimentally induced. By the mechanism of an anterograde alternation you prevent synaptic plasticity which should then also prevent the animal from memorizing or learning something. This is the most prominent method to prove

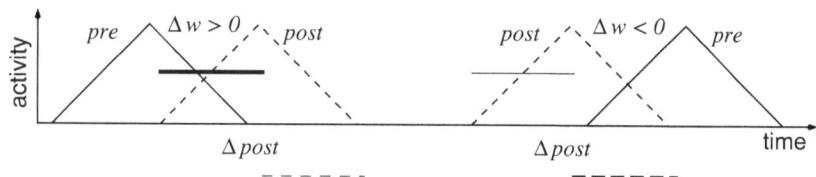

Figure 1.5: *Schematic diagram demonstrating STDP modeled by differential Hebbian plasticity ($\Delta w = pre \cdot \Delta post$) using a triangular shaped activity profile. If the pre-synaptic activity is before the post-synaptic one, the positive derivative is correlated with higher pre-synaptic activity. On contrary if the timing is inverted (second sequence), the higher pre-synaptic activity correlates more with the negative derivative. The horizontal bars are representing $\Delta post$ where a solid bar indicates a positive and dashed bar a negative value.*

the relation between plasticity and learning and one of the proofs was shown by Morris (1989) in his experiment. A rat had to find a platform within a circular box filled with water (Morris water maze). After blocking NMDA receptors the rat was no longer able to learn the location of the platform. Retrograde alternation implies that you could vary the synaptic strength in such a way that, for instance, a rat that learned the platform at a certain position could be 'reprogrammed' to find the platform at another location. This is also not yet experimentally possible. Hence so far, it has been shown in two ways that a relation between plasticity and learning exists.

Nonetheless, only a few theoretical learning rules can be directly related to the biophysics of Hebbian plasticity. Apart from the Hebbian learning rule one important example is the classical conditioning rule which we will discuss in the next section in the context of open-loop and closed-loop systems. The "open" means that there is no feedback to the system. In behaving systems the environment in which the system operates closes the loop leading to feedback. In this case we are talking about closed-loop systems.

1.2 Open-loop versus closed-loop learning using the example of classical and operant conditioning

In the previous section we discussed plasticity mostly at the synapse level and related plasticity to learning. One of the learning paradigms which is related to biophysical plasticity mechanisms is conditioning. In the late 19th century Pavlov (Pavlov, 1927) investigated the gastric function of dogs and recognized that his dogs were not only salivating when he presented food but often earlier. He found out that responses, e.g. the salivation, occur after stimuli which directly cause such a response, e.g. food; or after stimuli, e.g. a bell, which were repeatedly presented before the behavior-eliciting stimulus (see Figure 1.6 A). He called the stimulus which directly causes the response the unconditioned stimulus and the stimulus which was initially unrelated the

1.2 OPEN LOOP VERSUS CLOSED LOOP

conditioned stimulus. Besides this basic experiment Pavlov conducted different others. He, for instance, showed that dogs can unlearn the connection between the unconditioned and the conditioned stimulus if the latter was not followed by the first. This paradigm is called extinction. It is also possible to chain stimuli, i.e. the response shifts to a second unconditioned stimulus which was presented before the first. Secondary conditioning will become more important in chapter 4.

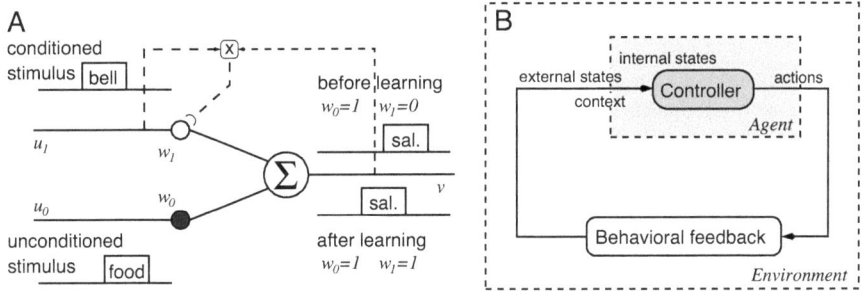

Figure 1.6: *Specific open-loop controller and schematic closed-loop system. In panel A we show a way to model classical conditioning with a correlation based learning rule. Here the bell is the conditioned stimulus u_0 and the food the unconditioned stimulus u_1. The response v represents the salivation of the dog, hence its action. In the course of learning the connection strength w_1 increases as both stimuli overlap. The connection w_0 between the unconditioned stimulus u_0 and the response v is fixed from the beginning as this stimulus needs to be sufficiently strong to elicit a response. In panel B an agent is embedded in its environment. By means of the controller the agent produces actions, which in turn influences the states the agent receives. The states usually consists of external and internal (e.g. memory) states and context information.*

All experiments described above were *open-loop* experiments. This means that Pavlov's dog had no influence on the behavior of Pavlov in particular on the presentation of the stimulus predicting food. Operant (or instrumental) conditioning was investigated in around the same time by Thorndike (1933) and Skinner (1933). The latter also coined the name for the Skinner box. In this box an animal, usually a rat, needs to press a lever to receive food. The opposite of such a confronting task is when the rat needs to avoid the lever in order to avoid receiving electric shocks. For the latter Porr and Wörgötter (2003a) extended the work on differential Hebbian learning. In their learning rule, called Isotropic sequence order (ISO) learning, the output activity influences the behavior of the system by interacting with the environment which in turn is responsible for the stimuli the system receives. As they closed the loop with the environment this is called *closed-loop* learning in contrast to open-loop paradigms like classical conditioning. A basic schematic is shown in Figure 1.6 B where an agent is embedded in its environment. As the loop is closed, the states the agent senses are related to the action it conducted. The controller could be, for instance, the diagram in panel A or rather a modification that is better suited for closed-loop operant conditioning.

In ISO learning the system's target is to avoid the conditioned stimulus which automatically evokes an unwanted response (or reflex). To this end, the system learns a temporal sequence of stimuli, i.e. it uses an earlier occurring stimulus to learn to predict the occurrence of a later stimulus. The resulting behavioral response then leads to the avoidance of the later stimulus. Additionally Porr et al. proved in Porr et al. (2003) that ISO learning by eliminating the later stimuli implements an inverse controller which is an important finding in an engineering sense.[2]

As Hebbian learning rules, including differential Hebbian learning, in general are not stable, i.e. the weight would, without additional mechanisms, increase to infinity, it makes more sense to investigate the properties of the different rules in a general way. For this we will develop a new theoretical framework (see chapter 2). However, closed-loop systems can guide us in the search and evaluation for useful, hence meaningful, plasticity rules and their parameters. For instance, a certain class of plasticity rules yields good results in the closed-loop paradigm of avoidance learning (see sections 2.1, 2.4, 2.5 and 3.1) and another class is more suitable for a goal-directed paradigm (see chapter 4).

1.3 Neuronal activity: Membrane potential, spikes or rates?

There are three different representations of neuronal activity: membrane potential (EPSP), spikes (action potentials), and rates (frequency). The first is the most accurate representation as it incorporates the complete time development of the membrane potential. In the next section we are, however, only interested in the timing of the spikes. As a consequence the representation becomes a point process and we are now speaking about temporal coding. To finally arrive at a rate code, we average over spikes in a given time window and take merely the number of spikes into account normalized by the width of this time window. Obviously there is no way back from rates to spikes. Starting from a temporal (spikes) code, there exists, however, a way to recover some aspects of the membrane potential. For this purpose each spike needs to be convolved with an EPSP kernel. These kernels (or filters) have various shapes. Most prominent are alpha functions, damped sine waves and difference of exponentials. The first was invented by Rall (1967) in order to describe EPSP at different potentials and is of the form $h(t) = \frac{t}{a}\exp(1-\frac{t}{a})\Theta(t)$. With the second, orthogonality among kernels is achieved by using different parameters (Porr and Wörgötter, 2003a, 2006). It writes as $h(t) = \frac{1}{b}\sin(bt)\exp(at)\Theta(t)$. The third (difference of exponentials) will be mainly used within this thesis because of its mathematical properties. It is given by:

$$h(t) = \frac{1}{\sigma}(e^{-at} - e^{-bt})\Theta(t). \qquad (1.1)$$

---

[2]It should be noted that the first relation of differential Hebbian learning to machine learning is by Kosco (1986) who examined features of differential Hebbian learning in the context of machine learning.

1.3 NEURONAL ACTIVITY

with $\Theta(t)$ being the Heaviside function and a, b, and σ being parameters that define the rising (a) and the falling (b) phase and the amplitude (σ) of the kernel. Actual parameters are given later, however, only $a < b$ results in positive values of h. As this kernel is used throughout this thesis, different shapes are plotted in Figure 1.7.

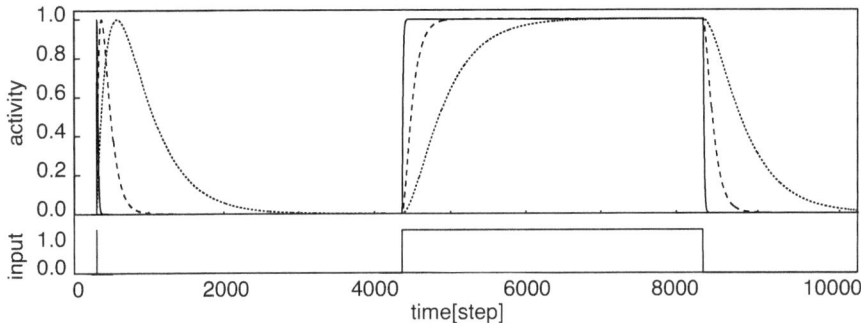

Figure 1.7: *Kernels with different parameters using equation 1.1 are shown. For the first group we convoluted the kernels with a delta peak and for the second group with a rectangular function (bottom). The dotted function spreads the input signal most strongly, however it also takes most time to reach its maximum. The parameters for the dotted functions are $a = 0.0025$, $b = 0.005$, and $\sigma = 0.25/200$ (for first and second signal, respectively). By contrast the solid function is shorter and faster. The parameters for the solid functions are $a = 0.01$, $b = 0.02$, and $\sigma = 0.25/5$. The dashed functions lies in between with parameters $a = 0.005$, $b = 0.01$, and $\sigma = 0.25/50$. A nice feature of this type of kernel is that the amplitude depends only on the ratio $\frac{a}{b}$. Therefore the amplitude for a ratio of 0.5 is $\frac{1}{4\sigma}$.*

Furthermore we note that when activity is spread out over time we need to provide some kind of memory mechanism without which individual events (spikes) can not be related to each other. This can be explained in the most basic way when discussing classical conditioning models (Figure 1.6 A). In order to learn to react to the earlier conditioned stimulus, it has to be *remembered* in the system. To this end, the concept of eligibility traces had been introduced (Hull, 1939, 1943; Klopf, 1972, 1982; Sutton, 1988; Singh and Sutton, 1996), where the synapses belonging to the earlier stimulus remain eligible for modification for some time until this trace fades. To implement such an eligibility trace one would need to convolute the stimuli with filters that spread out over time. In fact these filters are not different from the kernels used to emulate the EPSP except that they would need to cover seconds or minutes and not milliseconds. For simplicity we could just assume *one* process, thus one set of kernel parameters, which equally affects the neuron's output *and* its plasticity. Only in section 2.1 and section 4.2.3 we will discuss properties of different kernel processes.

1.4 Definitions and roadmap

In this thesis we will present all plasticity rules following the example of Figure 1.8. There the definitions for the symbols we will use throughout the text can be found as well. We use the kernel functions (equation 1.1) to convolve them with the input x_i. This will then be used for either the plasticity pathway alone or the plasticity and the output v pathway (see Figure 1.8 for the latter). In general a convolution is given by

$$(\xi * \eta)(t) = \int_0^\infty \xi(z)\, \eta(t-z)\, dz. \tag{1.2}$$

Additionally, we model a spike as a delta function $\delta(t-t_i)$ for spike time t_i, thus the convolution simplifies to a temporal shift in the kernel function h:

$$h(t - t_i) = \int_0^\infty \delta(t - t_i - z)\, h(z)\, dz. \tag{1.3}$$

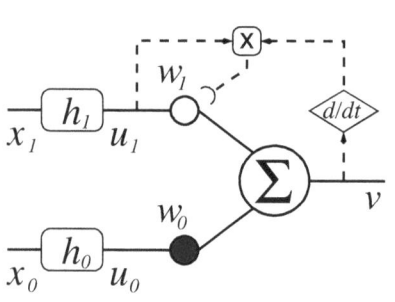

Symbol	Description
i	neuron
t	time
x_i	unfiltered input
u_i	filtered input
h_i	filter/kernel function
w_i	weight
$\dot{w}_i := \frac{d}{dt} w_i$	weight change
μ, α	plasticity/learning rate
v	output
r	reward
ac, cc	auto- and cross-correlation
G	functional

Figure 1.8: *Schematic diagram of a plasticity rule. The table describes the most important symbols used throughout this thesis. For a complete overview see list of symbols on page 129. The rounded box with the kernel function h describes a convolution (see equation 1.2) with the input x_i. The diamond-shaped box with the functional G which defines a mathematical operation using this functional. For instance, for differential Hebbian plasticity we need to set $G = d/dt$. The symbol Σ describes a linear summation of the inputs and the ×-symbol a multiplication. The solid lines are used for the output pathway and the dashed lines for the plasticity pathway. The semicircle at the end of a dashed line indicates a direct influence on the plasticity of a weight w.*

In chapter 2 we will first investigate the properties of differential Hebbian plasticity if only one synapse is plastic and all the others are kept fixed. This is done by using the theoretical framework of auto- and cross-correlations which we will describe in this chapter. The constraint

1.4 DEFINITIONS AND ROADMAP 11

of having only a single plastic synapse is lifted in chapter 3 where we derive analytical solutions for systems with many synapses. In chapter 4 we first introduce reinforcement learning which is similar to operant conditioning. Then we relate plasticity mechanisms to reinforcement learning and show three concrete realizations to asymptotically emulate temporal difference learning, which is a prominent reinforcement learning algorithm. The last chapter completes this thesis by concluding and discussing the results of this study and by providing an outlook including further ideas.

Chapter 2

Single-Plastic-Synapse Systems

In this chapter we will focus mainly on Hebbian plasticity $\dot{w}_i(t) = \mu\, u_i(t)\, v(t)$, in particular on differential Hebbian plasticity $\dot{w}_i(t) = \mu\, u_i(t)\, \dot{v}(t)$, and its mathematical description. The synaptic connection w_i changes through the correlation of pre-synaptic input u_i and post-synaptic output v. For the definition of the symbols see Figure 1.8. The underlying plasticity properties are partly used to investigate neuronal data but it is important to mention that all of the rules used here are at a much higher level of abstraction as compared to the biophysics of synapses. They can however be directly transfered to behaving systems.

In the following sections, sequences of two delta-pulses $x_{0/1}$ will be repetitively presented to the different systems, where x_1 comes earlier in time than x_0 with an interval of $T = t_{x_0} - t_{x_1}$ steps between them. The final weight change $\Delta\omega$ is calculated by integrating the respective learning rule: $\Delta\omega = \int_0^\infty \dot{w}(t)\, dt$ (see appendix B). From this the development of the weights can be plotted for multiple pulse pairs. In addition, we will investigate the different weight change curves plotting the weight change against the interval between inputs T. For negative T the temporal order of the pulses is inverted.

In general, plasticity is regulated by a plasticity rate (learning rate) which is usually below 1. In the following, we will use μ for the plasticity rate when talking about correlation-based learning and α for reinforcement learning. In this chapter the synaptic weight w will always be plotted in dimensions of the plasticity rate. Additionally, we demand a quasi-static or adiabatic condition, i.e. changes in synaptic strength are much smaller than the changes in the signals: $\frac{\dot{w}_i}{w_i} \ll \frac{\dot{u}_i}{u_i}$. This condition can be assured by setting the plasticity rate to values much smaller than 1: $\mu \ll 1$. This approach is commonly assumed for such systems (Dayan and Abbott, 2001) and it allows us to analytically calculate the weight change by neglecting the derivative of the weight w on the right hand side of the plasticity rule (see equation B.2). We also neglect the variability of the homogeneous solution for the calculation of the inhomogeneous part. For a detailed discussion on the differences which emerge when not using this assumption see appendix B.

We are especially interested in the stability of the plasticity rules. All rules considered here learn by cross-correlating two signals *with each other* (x_1 with x_0). Positive correlations of x_1 *with itself* (auto-correlations) are normally unwanted. As will be seen later, this leads to

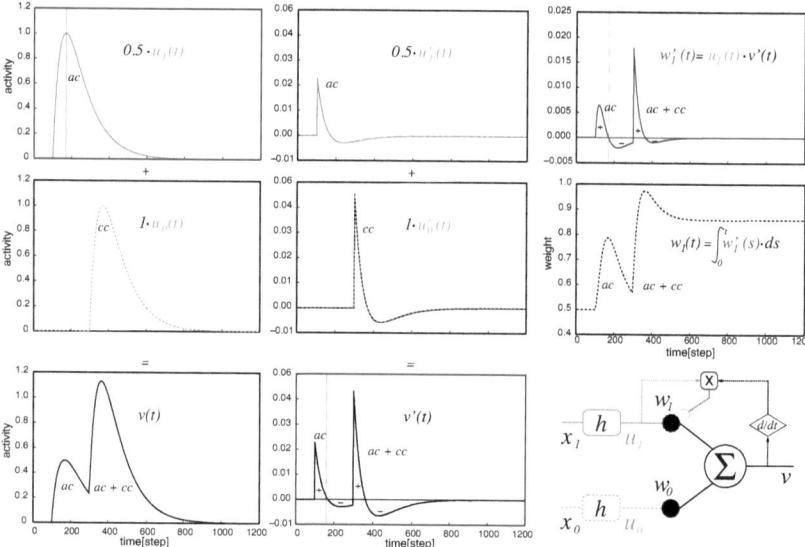

Figure 2.1: *Step by step explanation how the weight w_1 changes when using differential Hebbian plasticity (see section 2.1). In the bottom right corner, the architecture is shown where different paths have different shades of gray. These colors will be used for the signals, too. The left column shows the pure input signals and their weighted sum v which is the output. In the middle column, we see the derivative of the input signals and the output. The upper right panel shows the multiplication of the input signal u_1 and the derivative of the output, thus the derivative of the weight w_1 (darker gray panels, the dashed vertical line indicates the maximum of u_1). The integration of this panel is shown one panel below which, then, is the time development of the weight.*

weight divergence. Negative auto-correlations on the other hand act as a decay term which drive the synaptic weight to zero. Such leakage terms are commonly assumed in spiking neuron models (Gerstner and Kistler, 2002a). Hence to investigate these properties separately it makes sense to subdivide the contributions of the plasticity rule into a cross-correlation term Δw^{cc} and an auto-correlation term Δw^{ac} by: $\Delta w_1 = \Delta w_1^{ac} \cdot w_1 + \Delta w_1^{cc}$, the latter term drives the weight change of w_1 during the occurrence of x_0, whereas the auto-correlation term Δw_1^{ac} also changes the weight in the absence of the x_0 signal. Hence, the pure auto-correlation contribution becomes visible when switching x_0 off (see Figure 2.1 for a step by step example how the weight actually changes). If the auto-correlation is zero, this should stop weight change as there is no more cross-correlation existing. In the following diagrams we set $x_0 = 0$ at a certain time-step (mostly after 40 % of weight development, i.e. $t = 8000$) to show how auto-correlation influences the weight change for a given rule.

We will start with plain homosynaptic differential Hebbian plasticity comparing it to an older model and creating a hybrid version of both models. This chapter also includes an excursion which extends our investigations to homosynaptic Hebbian plasticity. The section covering temporal difference learning can be compared later with chapter 4. The analysis of the auto-correlation contribution will show that plain homosynaptic differential Hebbian plasticity (ISO learning) has unwanted (divergent) characteristics. Then we present two modifications that overcome this problem, namely heterosynaptic differential Hebbian plasticity (ICO learning) and homosynaptic Hebbian plasticity with a third factor (ISO3 learning), analyzing their properties in more detail.

2.1 Homosynaptic differential Hebbian plasticity - S&B model, ISO learning and VOT plasticity

The first model we investigate in more detail was designed by Sutton and Barto (1981). By presenting their model we break with our simplification that the plasticity and the output obey the same dynamics. In their original contribution they also use a different way to model these eligibility traces, namely a recursion. Although we start with their equations we will implicitly change to convolutions afterward. We will call their model S&B model. The synaptic weight change is governed by

$$\dot{w}_1(t) = \mu\, u_1(t)\, [v(t) - \overline{v}(t)], \qquad (2.1)$$

where they have introduced one eligibility trace at the input x_i and another at the output v given by:

$$u_1(t+1) = a_{\text{SB}}\, u_1(t) + x_1(t) \qquad (2.2)$$

$$\overline{v}(t+1) = b_{\text{SB}}\, \overline{v}(t) + (1 - b_{\text{SB}})\, v(t), \qquad (2.3)$$

with control parameters a_{SB} and b_{SB}. Mainly, they discuss the case of $b_{\text{SB}} = 0$ where $\overline{v}(t) = v(t-1)$ which results in the discrete form of a derivative: $\dot{v}(t)$. Thus their rule (Figure 2.2 A) turns into:

$$\dot{w}_1(t) = \mu\, u_1(t)\, [v(t) - v(t-1)] \qquad (2.4)$$
$$= \mu\, u_1(t)\, \dot{v}(t). \qquad (2.5)$$

This rule is *Hebbian* as the weight change is driven by a correlation of input and output and it is *differential* as not the output as such but its difference is taken into account. Furthermore, it is *homosynaptic* as weight w_1 changes due to the activity of the input connected to w_1, namely x_1. In section 2.4 we will discuss heterosynaptic plasticity where the activity of inputs not connected to the synapse under consideration drives the plasticity.

An important aspect of this rule was mentioned in the beginning which is the different dynamics for output and plasticity. Hence, the output needs to use either different kernel parameters (a_{SB}) or, even simpler, does not need to use any kernel at all:

$$v(t) = w_0 \cdot x_0(t) + w_1 \cdot x_1(t), \qquad (2.6)$$

Before learning, this neuron's output will only respond to the signal x_0, while after learning it will respond to x_1 as well.

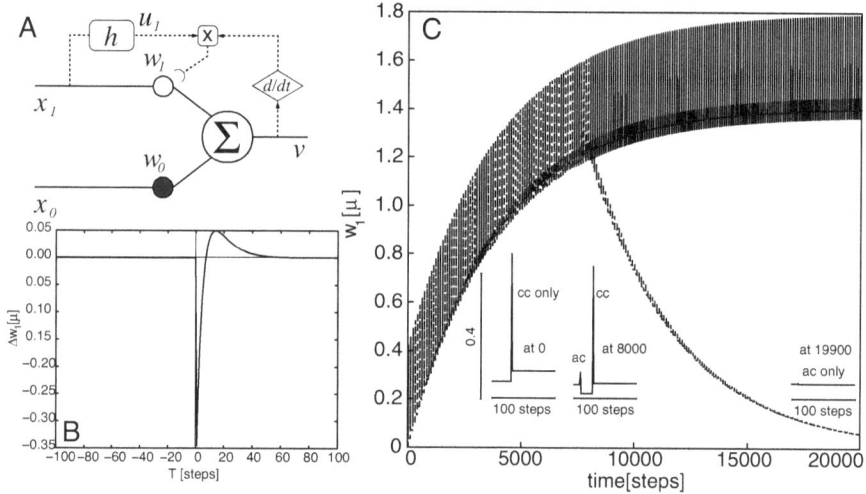

Figure 2.2: *Architecture and weight development of the S&B model. Panel A shows the architecture where only the plasticity path (dashed line) uses kernel functions. In panel B we plot the weight change for different timings of x_0 with respect to x_1 where a positive value of T means that x_0 is after x_1. Note that this curve only represents the cross-correlation part. Panel C shows an example of weight development in time of many x_1/x_0 pairs with (dashed) and without (solid) switching off the x_0 signal after time $t = 8000$. In the inset we plot a magnification of a single weight development step at certain times to show the difference between auto- and cross-correlation. Parameters were $w_0 = 1$, $a = 0.1$, $b = 0.2$, $\sigma = 0.25$, and $T = 20$.*

Let us now calculate the auto- and cross-correlation contributions for the S&B model, equation 2.5, when using spikes as inputs for $x_{0/1}$ at time $t = T$ and $t = 0$ respectively. Because

2.1 S&B MODEL, ISO LEARNING AND VOT PLASTICITY

we model spikes as a delta functions $\delta(t - t_i)$ for spike times t_i, the convolution simplifies to a temporal shift in the kernel function h (see equation 1.3). In a simplified way one writes

$$\begin{aligned}
\Delta w_1 &\cong \int_0^\infty u_1(t)\, \dot{v}(t)\, dt \\
&= \int_0^\infty u_1(t)\, \frac{d}{dt}(w_0 \cdot x_0(t) + w_1 \cdot x_1(t))\, dt \\
&\cong \int_0^\infty h(t)\, w_0\, \dot{\delta}(t - T)\, dt + \int_0^\infty h(t)\, w_1\, \dot{\delta}(t)\, dt \\
&= -\dot{h}(T)\, w_0 - \dot{h}(0)\, w_1
\end{aligned} \quad (2.7)$$

where we have assumed from line two to line three (indicated by the $\cong$) a quasi-static approach (see appendix B) and used $\int_0^\infty \dot{\delta}(t - t_0) f(t) = -\dot{f}(t_0)$ (Boykina, 2003). This gives us

$$\boxed{\Delta w_1^{ac} = -\dot{h}(0) \qquad \Delta w_1^{cc} = -w_0 \dot{h}(T)} \quad (2.8)$$

Note that the time derivative of the kernels used (e.g. equation 1.1) are always larger than zero at time $t = 0$. This leads to a negative auto-correlation of the S&B model, thus, to an intrinsic stability where the weight always drops to zero when no correlative signal is given. This is illustrated in Figure 2.2 C where we see the dashed curve converging to zero after x_0 was switched off. By contrast the solid curve develops asymptotically to the final weight which can be approximately calculated by $w_1^\infty = \frac{\Delta w_1^{cc}}{\Delta w_1^{ac}} = -\frac{w_0 \dot{h}(T)}{\dot{h}(0)}$ (see appendix G for more details). We will use this property when relating differential Hebbian plasticity to reinforcement learning in chapter 4. Additionally, one sees that the unfiltered input x_0 and its derivative lead to strong, needle-like excursions of the weight growth for every step, which let the line in the diagram appear broaden. These structures are caused by the cross-correlation part Δw^{cc} which is shown in the insets of Figure 2.2 C. The first close-up of the time development is at time $t = 0$, thus without any auto-correlation as weight w_1 is still zero. One can also see from these close-ups and from equation 2.8 that the amplitude is constant. These cross-correlation needle-like excursions disappear as soon as x_0 is switched off, however, the auto-correlation peaks are still there, decaying to zero in an exponential way. From equation 2.8, in particular Δw^{cc}, we also learn that the direction of plasticity, i.e. whether the weight converges against a positive or a negative value, depends on the phase of the kernel at the time input x_0. If the timing is inversed (x_0 before x_1, i.e. $T < 0$) the weight will not change at all. If the occurrence time of x_0 is before the maximum of the kernel response h to x_1, the final weight will be negative and only if x_0 occurs after the maximum of the kernel, weights will reach a positive value. This is summarized in Figure 2.2 B.

Assuming the same temporal characteristics for the plasticity and the output pathway, we arrive at the diagram Figure 2.3 A, which is called ISO learning (Porr and Wörgötter, 2003a). The ISO learning rule is identical to equation 2.5 from the S&B model, however, with a different output equation

$$v(t) = w_0 \cdot u_0(t) + w_1 \cdot u_1(t). \quad (2.9)$$

The weight change of a single signal pair for the ISO rule (equation 2.5 and equation 2.9) can be written as (see appendix D):

$$\Delta w_1^{ac} = \exp \int_0^\infty h(t)\dot{h}(t)dt - 1 = \exp \frac{1}{2}h^2(\infty) - 1 = 0 \qquad (2.10)$$

$$\Delta w_1^{cc} = w_0 \int_0^\infty h(t)\dot{h}(t-T)\,dt = \text{sign}(T)\,w_0\,\frac{b-a}{a+b}\frac{1}{2\sigma^2}h(|T|) \qquad (2.11)$$

where the auto-correlation term converges to zero for $t \to \infty$ as the kernels h eventually decay to zero.

Additionally, we calculate the time development of the cross-correlation part to give an insight into the exact weight change:

$$w_1^{cc}(t) = \frac{\Theta(t-T)\,\Theta(t)\,w_0}{2\,(a+b)\,\sigma^2}(-\text{sign}(T)\,\sigma\,(a-b)\,h(|T|)$$
$$- 2e^{-t(a+b)}(ae^{aT} + be^{bT})$$
$$+ (a+b)(e^{-a(2t-T)} + e^{-b(2t-T)})). \qquad (2.12)$$

Figure 2.3 C shows the step-by-step behavior of ISO learning. Weight growth is quite linear, however, due to a substantial numerical artifact, which we will discuss later, the weight increases exponentially indeed. This is also the reason why, after switching off x_0, weights will drift upwards. This drift decreases for very small integration step sizes Δt and large relaxation times t.

The bottom insets of Figure 2.3 C show the relaxation behavior of the weight for a single input pulse pair at different times. At time $t = 0$ only the cross-correlation part is visible whereas at time $t = 8000$ an early auto-correlation component is followed by a big, cross-correlation dominated hump as soon as x_0 occurs (see also Figure 2.1 for a detail example). The curve relaxes to the final weight value after some time depending on the filter characteristic of h. In the insets at the upper left of panel C, we compare the auto-correlation component of the weight change for narrow and wide kernel functions. The right curve shows that, following equation 2.10, the auto-correlation indeed approaches zero for wider kernels and longer times. This is different for the left curve which represents the auto-correlation contribution when using coarser sampling. Here we see a potentially very strong source of error: The auto-correlation contribution does not vanish anymore. This is a purely numerical artifact of the integration procedure (see appendix C for a further discussion), but - as high sampling rates are often too costly (for example in real-time applications) - this artifact can strongly interfere with the convergence of ISO learning. Hence, we are facing two potential sources of error: (1) The tardy relaxation behavior of (essentially) the cross-correlation term (insets of Figure 2.3 panel C). This error becomes relevant when pulse pairs follow each other in time too quickly. And (2) the non-negligible numerical error that renders the auto-correlation to non-zero values even for long relaxation times. In this and the following sections we will discuss other differential Hebbian rules which have been invented to solve these problems.

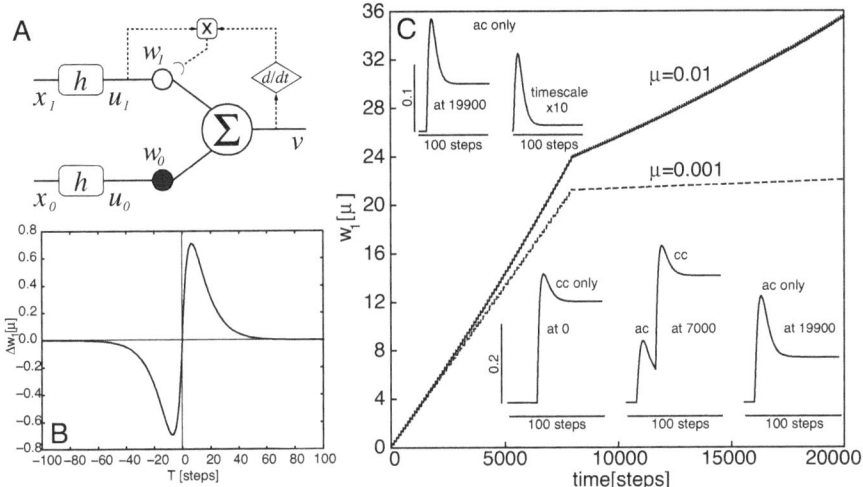

Figure 2.3: *Architecture and weight development of ISO learning. Panel A shows the architecture where both paths use the same kernel function. In panel B, we plot the weight change for different timings of x_0 with respect to x_1 where a positive value of T means that x_0 is after x_1. As the auto-correlation is ideally zero, this curve represents the whole weight change. Panel C shows an example of weight development in time of many x_1/x_0 pairs with (dashed) and without (solid) switching off the x_0 signal after time $t = 8000$. In the inset we plot a magnification of a single weight development step at certain times to show the difference between auto- and cross-correlation, and, additionally, the difference between different time scales (upper left part). Parameters were $w_0 = 1$, $a = 0.1$, $b = 0.2$, $\sigma = 0.25$, and $T = 20$.*

The weight change curve (Figure 2.3 B) of ISO learning is anti-symmetrical (Porr and Wörgötter, 2003a). As long as kernels for input and output are the same, this curve will have identical shapes on both sides (Figure 2.3 B). This is interesting, because with this rule a completely isotropic setup can be designed, in which both synapses are allowed to change as will be discussed later when investigating multi-synapse systems (see section 3.1 and section 3.2).

In the S&B model we used delta functions for the output (equation 2.6) and in ISO learning the same kernel functions as in the plasticity pathway (equation 2.9). In order to generalize we define the output as

$$v(t) = w_0 \cdot (x_0 * h_v)(t) + w_1 \cdot (x_1 * h_v)(t)$$
$$= w_0 \cdot u_{v,0}(t) + w_1 \cdot u_{v,1}(t) \qquad (2.13)$$

where we indicate different parameter values a_v, b_v and σ_v of the kernel function h_v (equation 1.1) with an index v. Figure 2.4 A shows the architecture of the rule which in the following we will refer to as Variable Output Trace (VOT) plasticity as it uses variable output traces, which leads to different time scales for plasticity and output. In the limit of a_v and b_v to infinity, this model resembles the S&B model, and for $a_v = a$ and $b_v = b$, ISO learning The calculation of the weight change results now in

$$\Delta w_1^{ac} = \frac{(a-b)(a_v - b_v)(ab - a_v b_v)}{\sigma^2(a + a_v)(a_v + b)(a + b_v)(b + b_v)} \quad (2.14)$$

$$\Delta w_1^{cc} = -\frac{(a_\xi - b_\eta) w_0}{\text{sign}\, T\, \sigma^2} \left(\frac{a_\eta\, e^{-a_\eta |T|}}{(a_\eta + a_\xi)(a_\eta + b_\xi)} - \frac{b_\eta\, e^{-b_\eta |T|}}{(b_\xi + b_\eta)(a_\xi + b_\eta)} \right) \quad (2.15)$$

where $\xi = v, \eta = \emptyset$ if $T \geq 0$ and $\xi = \emptyset, \eta = v$ if $T < 0$

where ø indicates that no index needs to be used (e.g. $a_\emptyset = a$).

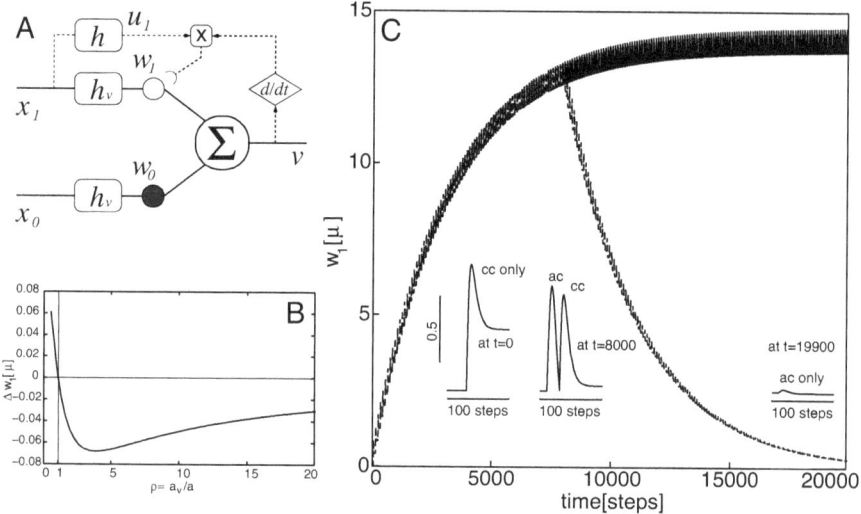

Figure 2.4: Architecture and weight development of a model with different time scales (VOT plasticity). Panel A shows the architecture where the plasticity path (dashed line) uses kernel functions different from the output pathway (solid line). In panel B we plot the weight change for different ratios of $\frac{a_v}{a}$ by varying the output trace. Note that this curve only represents the auto-correlation part and is independent of T. Panel C shows an example of weight development in time of many x_1/x_0 pairs with (dashed) and without (solid) switching off the x_0 signal after time $t = 8000$. In the inset, we plot a magnification of a single weight development step at certain times to show the difference between auto- and cross-correlation. Parameters were $w_0 = 1$, $a = 0.1$, $b = 0.2$, $\sigma = 0.25$, $\rho = 5$, and $T = 20$.

In order to have a weight decay, Δw^{ac} needs to be negative. As our parameters (a_η, b_η, and σ_η) are strictly positive the denominator is positive, too. Further, as we need $a_\eta < b_\eta$ for

2.1 S&B MODEL, ISO LEARNING AND VOT PLASTICITY

positive values of h, the first two terms in the numerator are negative, however, their product is positive. Therefore, only the last term will decide whether the weight change of the auto-correlation is negative or not. When assuming a certain relation $\Upsilon_\eta = \frac{b_\eta}{a_\eta}$, which needs to be strictly larger than one, we get $\Upsilon a^2 - \Upsilon_v a_v^2 < 0$ which gives us a condition for a_v in relation to a to achieve negative auto-correlations:

$$\rho := \frac{a_v}{a} > \sqrt{\frac{\Upsilon}{\Upsilon_v}} \qquad (2.16)$$

If Υ_v is of the same order as Υ, we find that it suffices for the output kernels to have parameters a_v and b_v larger than the plasticity kernels, i.e. the output pathway needs to have a shorter time scale than the plasticity pathway[1]. This is shown in Figure 2.4 B, where we plot the auto-correlation part of the weight change for different ratios of $\frac{a_v}{a}$. Ratios larger than 1 produce negative auto-correlations, so that such systems are convergent, and ratios smaller than 1 produce positive auto-correlations, which leads to divergent systems.

For the weight development in Figure 2.4 C, we set $\rho = 5$, which gives us still needle-like excursions, however, not as pronounced as in the S&B model. This also shows that the decay is adjustable by means of the ratio. In the close-ups we again find the two separate phases of plasticity which are governed by the two inputs $x_{0/1}$ of the output which has a smaller time scale than the plasticity kernel for u_1.

Additionally, we show in panel B of Figure 2.4 the weight change curves of the cross-correlation part for different ratios of $\frac{a_v}{a}$. The zero-crossing (zero weight change) shifts from zero at $\rho = 1$ (ISO learning) to positive values $\rho \to \infty$. At infinity we would find that the weight change for $T < 0$ has vanished (S&B model).

We note that biologically realistic neuron models commonly use different time scales for output and plasticity. Such models where usually the shape of the kernels is varied (Saudargiene et al., 2004) are used to describe site-specific plasticity (Saudargiene et al., 2005; Tamosiunaite et al., 2007), hence plasticity which is different for different locations of synapses on a dendrite.

Next we investigate the question of convergence. When do all these different algorithms converge? Trivially, weight growth at w_1 will stop as soon as $x_1 = 0$ in all cases. Theoretically, plasticity rules with identical time scales like ISO learning converge as soon as the second signal x_0 vanishes. This corresponds to the fact that the auto-correlation is zero. However, as discussed, this particular plasticity rule is highly sensitive to errors, which can easily destroy convergence. Additionally, we find that weights will converge if $T = 0$ (see Figure 2.3 B). Hence these systems will be essentially stable if small positive values of T are followed by small negative ones (or vice versa). For plasticity rules with negative auto-correlations, e.g. VOT plasticity, weights convergence as soon as equation 2.16 is fulfilled. Namely, weights will either reach

$$w_1^\infty = \frac{\Delta w_1^{cc}}{|\Delta w_1^{ac}|} \qquad (2.17)$$

if both signals x_0 and x_1 are existent, or zero if only x_1 is given.

---
[1]For instance, to achieve a delta-function like in the S&B model, a_v needs to reach infinity.

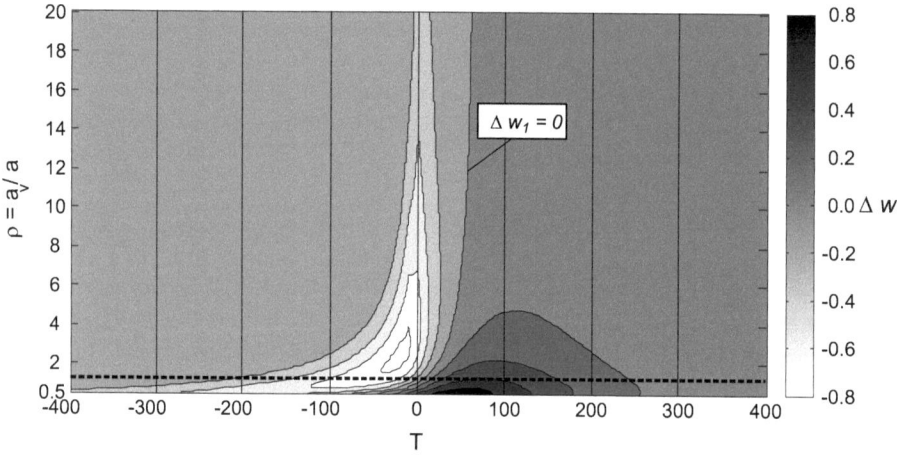

Figure 2.5: *Weight change curves of the cross-correlation part for different value ratios ρ and T values for VOT plasticity. Colors indicate different contributions of the cross-correlation. The zero-crossing shifts from $T = 0$ at $\rho = 1$ which resembles ISO learning (dashed line) to $T > 0$ for $\rho \to \infty$. At infinity, we would find that the weight change for $T < 0$ has vanished which corresponds to the S&B model. Parameters were $w_0 = 1$, $a = 0.01$, $b = 0.02$, $\sigma = 0.25$.*

2.2 Homosynaptic plain Hebbian plasticity - Hebb learning

Although we are mainly focusing on differential Hebbian plasticity in this thesis, we would like to take a short glance at plain Hebbian plasticity here (Gerstner and Kistler, 2002b). The only difference to differential Hebbian plasticity is that we use the plain output of the neuron instead of its derivative. The rule is schematically shown in Figure 2.6 A and writes as

$$\dot{w}_1(t) = \mu \, u_1(t) \, v(t). \tag{2.18}$$

where the output is the weighted sum of both inputs (equation 2.9).

The auto- and cross-correlation contributions of Hebbian plasticity are calculated to

$$\Delta w_1^{ac} = \frac{(\alpha - \beta)^2}{2\,\alpha\,\beta\,(\alpha + \beta)\,\sigma^2} \qquad \Delta w_1^{cc} = w_0 \frac{\alpha - \beta}{2\,(\alpha + \beta)\,\sigma} H(|T|) \tag{2.19}$$

where $H(t)$ is the antiderivative of $h(t)$. As the difference in the numerator is squared, the auto-correlation is strictly positive and always causes weight divergence to infinity. This is shown in Figure 2.6 C, where it does not really matter anymore whether there is a cross-correlation or not. The steepness of both (with and without x_0) curves representing the weight development is similar and weights diverge. The weight change curve in Figure 2.6 B is not temporally

2.3 TD LEARNING

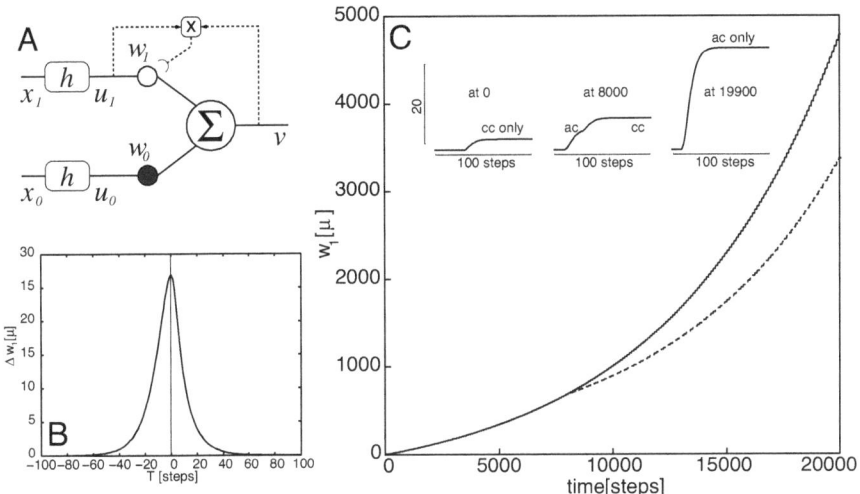

Figure 2.6: *Architecture and weight development of Hebbian plasticity. Panel A shows the architecture where both paths use the same kernel function. In panel B, we plot the weight change for different timings of x_0 with respect to x_1 where a positive value of T means that x_0 is after x_1. Note that this curve only represents the cross-correlation part. Panel C shows an example of weight development in time of many x_1/x_0 pairs with (dashed) and without (solid) switching off the x_0 signal after time $t = 8000$. In the inset, we plot a magnification of a single weight development step at certain times to show the difference between auto- and cross-correlation. Parameters were $w_0 = 1$, $a = 0.1$, $b = 0.2$, $\sigma = 0.25$, and $T = 10$. Note as the positive auto-correlation causes the system to self-amplify, the plasticity rate needs to be a small number. Here, we set it to $\mu = 0.001$.*

asymmetrical anymore, thus, independent of whether x_1 comes before or after x_0, the weight changes only according to the absolute value of T. The maximal weight change is achieved when both signals occur at the same time.

2.3 Temporal difference learning - TD learning

TD learning (Sutton, 1988) was developed a few years after the S&B model (Sutton and Barto, 1981) in order to have a model for the prediction of delayed rewards. One centrally new aspect at that point had been to introduce a reinforcement signal r, which affects the plasticity, but not the output of the system (Figure 2.7 A). The synaptic weight changes according to:

$$\dot{w}_1(t) = \mu\, u_1(t)\, [r(t) + \gamma\, v(t+1) - v(t)]$$
$$\approx \mu\, u_1(t)\, [r(t) + \dot{v}(t)]$$
$$\approx \mu\, u_1(t)\, \delta_r(t). \qquad (2.20)$$

where we define
$$\delta_r(t) = r(t) + \gamma\, v(t+1) - v(t) \qquad (2.21)$$
as the δ error of TD learning, which is the mismatch between predicted (expected) and actual reward. The parameter $\gamma \leq 1$ is called the discount factor which accounts for the fact that distant rewards should usually be valued less. For simplicity we set it here to $\gamma = 1$, thus having no discounting. Additionally we set $r = \tilde{r}\,\delta(t-T)$, where $\tilde{r}$ is the reward amplitude.

Figure 2.7 C shows the temporal development of the weight change for TD learning. Note in TD the reward does not enter into the output of the neuron, but only influences the learning and the rule becomes identical to the S&B rule if we remove r or x_0, respectively. To this end we also investigate the properties of TD learning at $r = x_0 = 0$.

The auto- and cross-correlation contributions of the TD learning rule (equation 2.20) look similar to the contributions of the S&B model:

$$\Delta w_1^{ac} = -\dot{h}(0) \qquad\qquad \Delta w_1^{cc} = +\tilde{r}\, h(T) \qquad (2.22)$$

The only difference is a changed sign and the derivative of the kernel h associated with the x_0, or rather r, signal. The reason for the missing derivative at the cross-correlation kernels is that the r signal enters the plasticity pathway directly and does not take the detour through the output where the derivative comes from.

Therefore in TD learning (Figure 2.7 C), weights grow about ten times faster than in the S&B model (Figure 2.2). When we switch r (or x_0) off, we find in both cases that w_1 drops in the same way. Note, this is not what ought to be done in a TD rule. Switching off δ_r would be the appropriate convergence condition,[2] and obviously weights will be - by construction - stable then. Still, as mentioned before, we want to look at the $r = 0$ case, because it directly corresponds to the $x_0 = 0$ condition of the differential Hebbian plasticity rules and shows the behavior of the pure auto-correlation contribution.

TD learning produces an asymmetrical weight change curve (Figure 2.7 B). This is a consequence of the missing kernel in the reward pathway. If early, the reward has already vanished before x_1 occurs and the correlation result remains zero.

For TD learning there exist two different types of convergence. The first one occurs at the end of our temporal development (Figure 2.7 C) where, similar to the S&B model, weights converge according to equation 2.17 with $w_1^\infty = \frac{\Delta w_1^{cc}}{\Delta w_1^{ac}} = -\frac{\tilde{r}\, h(T)}{\dot{h}(0)}$. It is interesting that this does not correspond to the condition $\delta_r(t) = 0$. This difference emerges from the fact that we use different kernels or rather eligibility traces compared to the original idea proposed by Sutton (1988). There, constant ($\gamma = 1$) or exponentially decaying ($\gamma < 1$) eligibility traces

---
[2] If $\delta_r = 0$ then the output v correctly predicted upcoming rewards

2.3 TD LEARNING

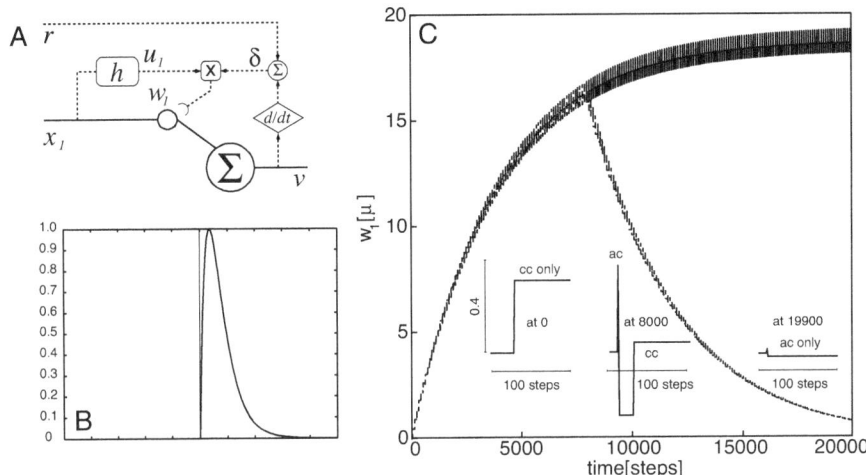

Figure 2.7: *Architecture and weight development of temporal difference learning. Panel A shows the architecture where only the plasticity path (dashed line) uses kernel functions. In panel B we plot the weight change for different timings of r with respect to x_1, where a positive value of T means that r is after x_1. Note that this curve only represents the cross-correlation part. Panel C shows an example of weight development in time of many x_1/r pairs with (dashed) and without (solid) switching off the r signal after time $t = 8000$. In the inset, we plot a magnification of a single weight development step at certain times to show the difference between auto- and cross-correlation. Parameters were $w_0 = 1$, $a = 0.1$, $b = 0.2$, $\sigma = 0.25$, and $T = 10$.*

with an amplitude of $\gamma^t h$ at time t have been applied. If we change the kernel functions to the kernels used in Sutton (1988), we can calculate the value to which the weight will converge in a simple way. For this we need to determine the following expressions: $\dot{h}(0)$ and $h(T)$ (see above). To get the correct value $\dot{h}(0)$, we need to take the positive limit of this derivative: $\lim_{t \to 0^+} \dot{h}(t) = -h(0) = -h$ and the value $h(T)$ which depends on the discount factor γ and the temporal difference T: $h(T) = \gamma^T h$. This causes the weight to converge to $w_1^\infty = -\frac{\tilde{r} h \gamma^T}{-h} = \gamma^T \tilde{r}$, which is the intended solution of temporal difference learning. However, even here the weight will not be stable but change during the occurrence of a x_1/r pair but in such a way that both contributions sum up to zero. This is shown in the inset of Figure 2.7 C where the absolute change of the weight is much higher than the individual changes at the beginning and the end of the pulse pair (at $t = 8000$).

The second way that convergence can be guaranteed concerns the δ error as such and, obviously, as soon as the δ_r signal is zero, weight w_1 will not change anymore. It has been long discussed that TD learning could be related to dopaminergic responses in the brain. Especially the behavior of some cells in the substantia nigra and ventral tegmental area (VTA) suggest that they represent the δ error of TD learning. Models which behave in a similar way have been made by Suri and co-workers (Montague et al., 1996; Suri and Schultz, 1998, 1999, 2001; Suri et al., 2001). The δ error decays to zero during learning as the negative derivative of the output always cancels out with the reward. This is shown in Figure 2.8, where a filter that stops when the reward occurs was used. Usually the occurrence of the reward is not known and to overcome this problem you need to employ more weights and a serial compound representation (see panel C of Figure 2.8) which is similar to a bank of kernels (see section 3.1). The problem with these models, however, is that it is difficult to find appropriate biophysical equivalents for the implementation of the TD rule.

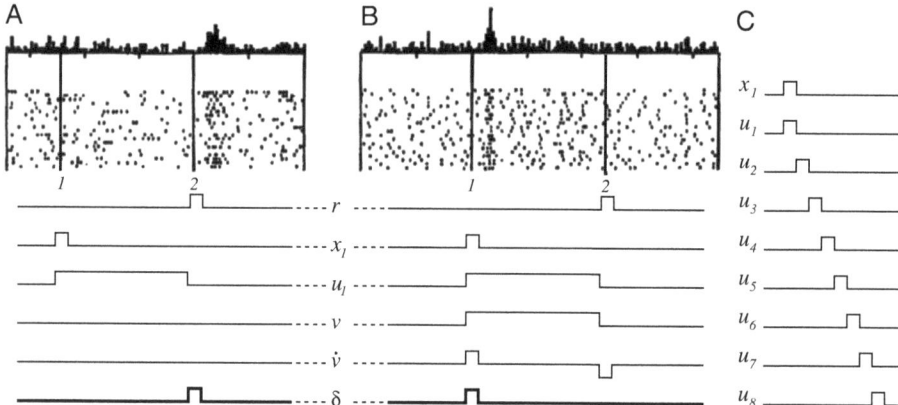

Figure 2.8: *Neuronal response in the basal ganglia showing the firing of a Dopamine neuron before (A) and after (B) learning. Both panels show in the upper part the spike patterns of a single neuron. The lower part shows the signal structure according to temporal difference learning which tries to connect the δ error with the Dopamine response. Panel A shows the Dopamine response and the signals before learning. The reward was not expected and therefore the δ error is nonzero shortly after (about 100 ms delay) the Dopamine neuron fires. After learning (panel B), the output predicts the reward and therefore the Dopamine neuron does not fire and the δ error stays zero at time point 2. However, the δ error and also the Dopamine firing shift before time point 1, which leads to no correlation between the u_1 signal and the δ error, thus, learning stops. Note that for $\dot{v}$ we used the difference between $v(t+1)$ and $v(t)$. This figure was recompiled from Schultz et al. (1997). In panel C, we depict a serial compound representation. Each of the filtered inputs u_i is delayed by $\Delta t = i - 1$.*

Note, for the $\delta_r = 0$ condition, the output v, or rather its derivative, needs to take on a certain value as opposed to ISO learning, where the input x_0 needs to become zero. Hence, we have TD learning where convergence is guaranteed by output-control as opposed to ISO learning, which uses input-control to guarantee convergence. Looking back at Figure 2.7 C, it becomes clear that setting $r = 0$ does not enforce convergence to a positive weight value. Setting $r = 0$

was only done for auto-correlation term evaluation. Furthermore, we note that there is no generic way to rephrase the TD-specific convergence criterion $\delta_r = 0$ into an input convergence condition. In order to attempt this, we would have to use input terms $(x_{0/1}, r)$ only, which can not be achieved.

2.4 Heterosynaptic differential Hebbian plasticity - ICO learning

In order to address the above stated problem about the sensitivity to numerical errors, we need to design a plasticity rule for which the auto-correlation term truly vanishes. For this, we modified the ISO learning rule in the following way.

Figure 2.9 A shows an architecture where we have replaced the derivative of the output in ISO learning with the derivative of the (later) input x_0. Hence, we are only correlating inputs with each other, thus, the name of this rule: Input correlation learning (ICO, Porr and Wörgötter, 2006). The plasticity rule is given by

$$\dot{w}_1(t) = \mu\, u_1(t)\, \dot{u}_0(t) \qquad (2.23)$$

and, as output and plasticity use the same kernels, the output follows equation 2.9 of ISO learning. Note that the output is not needed to drive plasticity.

Apart from the missing w_0 equation 2.23 looks exactly like the cross-correlation term of equation 2.5. Thus the overall weight change with this exception is identical to equation 2.11. We note that the auto-correlation is not existent. This is summarized by the following equations:

$$\Delta w_1^{ac} \equiv 0 \qquad \Delta w_1^{cc} = \frac{b-a}{a+b}\frac{\operatorname{sign} T}{2\sigma^2}\, h(|T|). \qquad (2.24)$$

The corresponding results are shown in Figure 2.9 B,C. The learning window is identical to that of the ISO rule (Figure 2.9 B), but now weights are stable for $x_0 = 0$. The insets in (C) show the relaxation behavior for a single pulse-pair. In comparison to ISO learning (inset in Figure 2.3 C) the shallow initial rising phase is missing here as there is no auto-correlation contribution. For the same reason the hump is a little bit smaller. This effect, however, is barely visible even when we would overlay the curves. Incidentally, ICO is identical to ISO in the limit of $\mu \to 0$. The ICO rule has been proven to be very useful in difficult learning tasks (Porr and Wörgötter, 2006). In fact this rule reliably works even with very high learning rates and will always converge if one manages to bring x_0 down to zero.

One should, however, notice that ICO learning is a form of non-Hebbian (heterosynaptic) plasticity, which may be less realistic from a biological point of view. Such heterosynaptic learning was only found at a few specialized synapses (Humeau et al., 2003; Tsukamoto et al., 2003). Heterosynaptic plasticity is usually associated with modulatory processes and not directly with Hebbian plasticity.

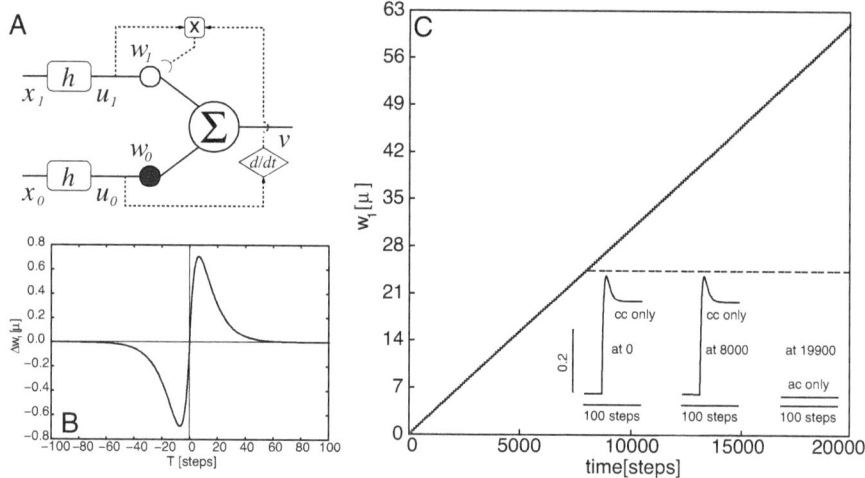

Figure 2.9: *Architecture and weight development of ICO learning. Panel A shows the architecture where both paths use the same kernel function. In panel B we plot the weight change for different timings of x_0 with respect to x_1, where a positive value of T means that x_0 is after x_1. As the auto-correlation is zero, this curve represents the whole weight change. Panel C shows an example of weight development in time of many x_1/x_0 pairs with (dashed) and without (solid) switching off the x_0 signal after time $t = 8000$. In the inset, we plot a magnification of a single weight development step at certain times to show the difference between auto- and cross-correlation. Parameters were $w_0 = 1$, $a = 0.1$, $b = 0.2$, $\sigma = 0.25$, and $T = 20$.*

2.5 Homosynaptic differential Hebbian plasticity with a third factor - ISO3 learning

ICO learning is very stable but, as mentioned above, it is a form of non-Hebbian plasticity, where the output does not influence the learning. This may be undesirable in certain cases. Therefore, efforts have been made to stabilize the ISO learning rule (Porr and Wörgötter, 2007). This can be achieved using a third factor, which has been called the "relevance signal" R (Figure 2.10 A). For practical purposes, most of the time we set it equal to x_0, but one should realize that - like the reward line in TD learning - R is indeed an independent signal. The signal R is meant to arise when for the animal/agent a behaviorally relevant event occurs.

The learning rule is similar to the ISO rule (equation 2.5):

$$\dot{w}_1(t) = \mu\, u_1(t)\, \dot{v}(t) \bar{R}(t) \qquad (2.25)$$

2.5 ISO3 LEARNING

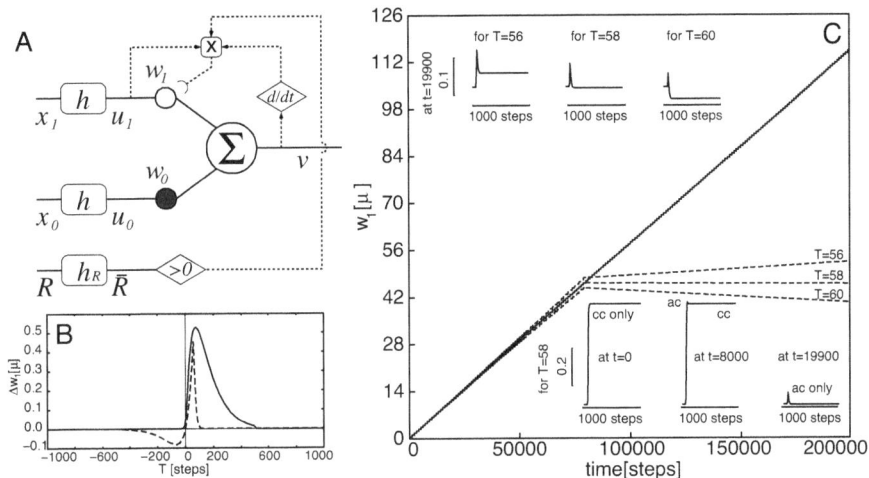

Figure 2.10: *Architecture and weight development of ISO3 learning. Panel A shows the architecture, where both paths use the same kernel function. In panel B we plot the weight change for different timings of x_0 with respect to x_1, where a positive value of T means that x_0 is after x_1. Note that this curve only represents the cross-correlation part for $T_R = T$ (solid) and $T_R = 58$ time steps (dashed). See Figure 2.12 for all possible values of T_R. Panel C shows an example of weight development in time of many x_1/x_0 pairs with (dashed) and without (solid) switching off the x_0 signal after time $t = 80000$. We chose $T = 58$ on purpose as this value brings the auto-correlation contribution closest to zero. Therefore we additionally plot the development for T values with ± 2 time steps. In the inset we plot a magnification of a single weight development step at certain times to show the difference of the auto-correlation part between the different T values (upper left part). Parameters were $w_0 = 1$, $a = 0.01$, $b = 0.02$, $a_R = 0.1$, $b_R = 0.2$ and $\sigma = 0.25$. Note that we needed to broaden the kernels to find a T value with sufficient small auto-correlation contribution.*

with an additional factor $\bar{R}$ which is the filtered version of R with kernel h_{a_R,b_R}. The weight change is:

$$\Delta w_1^{ac} = \int_0^\infty h(t)\dot{h}(t)h_{a_R,b_R}(t-T_R)dt \qquad (2.26)$$

$$\Delta w_1^{cc} = w_0 \int_0^\infty h(t)\dot{h}(t-T)h_{a_R,b_R}(t-T_R)dt \qquad (2.27)$$

where we introduced a new time interval T_R, which regulates the timing of the third factor. It is interesting that the auto-correlation term for ISO3 is, in general, unequal to zero, when using a single plastic synapse pointing to a possible instability. This is shown in Figure 2.11 B.

To explain the optimal position of T, Figure 2.11 A shows the signal structure. Let us assume that u_1 reaches its maximum exactly at T. As $\dot{v}(t) = \dot{u}_1(t)$ for $t < T$ and $T > 0$, we

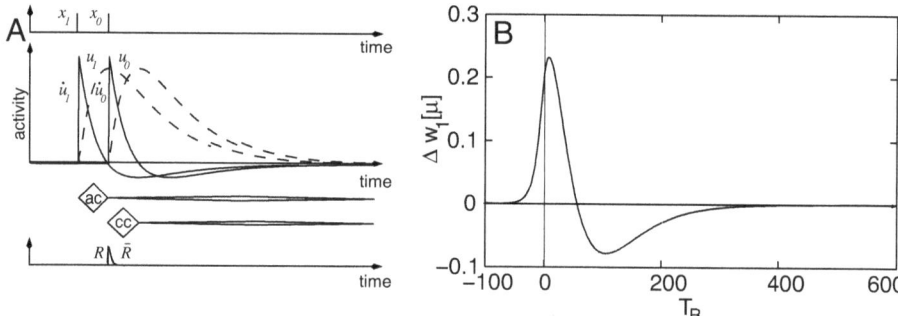

Figure 2.11: *Sketch of the signal structure if the auto-correlations of ISO3 learning is zero and weight change plot for the auto-correlation part of ISO3 learning. In panel A the timing is chosen in such a way that the auto-correlation contribution does not exist as the third factor occurs at the time when the derivative of the auto-correlation part is zero. Panel B shows the contribution of the auto-correlation for different values of T_R. If the relevance signal R had been a delta peak, the weight change for $T_R < 0$ would be zero. Parameters were $w_0 = 1$, $a = 0.01$, $b = 0.02$, $a_R = 0.1$, $b_R = 0.2$, $\sigma = 0.25$.*

have $\lim_{t \to T_-} \dot{v}(t) = \dot{u}_1(T) = 0$. This is the situation for the time-development in Figure 2.10 C when setting $T = 58$ time steps. If we furthermore assume that the R signal is very short (e.g. using a delta-pulse for R) and that it also happens at T, then learning only takes place at this moment in time, hence for $t = T$. As $\dot{u}_1(T) = 0$ we have totally eliminated the auto-correlation contribution. The outcome of panel C in Figure 2.10 is obtained under this condition and ISO3 is stable (compare the close-up signal structures of ISO in the inset of Figure 2.3 C and ISO3 in the inset of Figure 2.10 C). The insets show the relaxation behavior of ISO3, which is more like a step, for a single pulse-pair in comparison to ISO, which is curved. This demonstrates instantaneous relaxation of ISO3. Clearly, this example is constructed as T is usually unknown such that $\lim_{t \to T_-} \dot{v}(t) = 0$ can not be generally assured. This is shown in Figure 2.10 C, where we vary T by only ±2 time steps. Also the weight change curve changes with different values of T_R. Figure 2.10 B only shows it for $T = T_R$, but for different values of T_R we receive many learning curves, which is shown in Figure 2.12. It reveals that the zero crossing (zero weight change) moves along $T \approx T_R + 58$ for positive values of T_R and along $T \approx -58$ for negative T_R values. The shift of ~ 58 corresponds to the T_R, for which the auto-correlation is zero (compare to Figure 2.11).

Due to the problem that the situation in Figure 2.11 can not be generally assured (unknown T), it seems we have not gained anything so far by introducing ISO3. However, the situation changes when using a kernel bank to spread signal x_1 out in time (see section 3.1). Then, one can prove that the condition $\lim_{t \to T_-} \dot{v}(t) = 0$ will self-emerge as a consequence of the learning when using enough kernels (see section 3.1). Thus, when using a kernel bank, ISO3 becomes

2.6 DISCUSSION

a very stable method, indeed. Obviously, when coupling the relevance signal R with input x_0, weight development will also be stable as soon as x_0 is switched off.

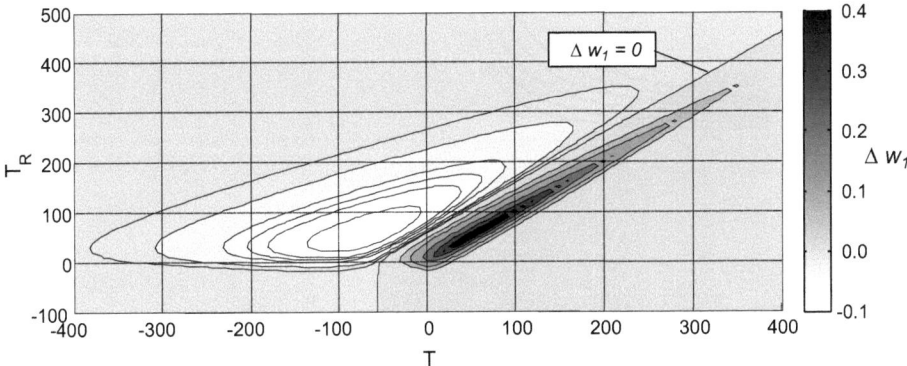

Figure 2.12: Weight change plot of the cross-correlation part for different time intervals T and T_R. Colors indicate different contributions of the cross-correlation. The zero crossing moves along $T \approx T_R + 58$ for positive values of T_R and along $T \approx -58$ for negative T_R values. Parameters were $w_0 = 1$, $a = 0.01$, $b = 0.02$, $a_R = 0.1$, $b_R = 0.2$, $\sigma = 0.25$.

2.6 Discussion

The here developed framework based on auto- and cross-correlation terms made it possible to compare different plasticity rules in a coherent way. The analysis so far has revealed several important common aspects.

Overall weight development

The final weight of plasticity rules without an auto-correlation contribution (ISO/ICO learning) is just the sum of all single weight change contributions with a given time delay T. Hence, in general the analyzed rules are linear: If the weight change curve for all given temporal differences T is known and all the temporal differences which will occur in the future are known, too, the final weight can be determined to $w_1^\infty = \sum_n \Delta w_1^{cc}(T_n)$. On the other hand if the auto-correlation contribution is negative (S&B model, VOT plasticity for $\rho < 1$), the overall weight development follows the difference equation G.1 and the final weight is calculated $w_1^\infty = \frac{\Delta w_1^{cc}(T)}{|\Delta w_1^{ac}|}$ (see equation 2.17). However, this is only true for constant T values as the cross-correlation depends on temporal difference T between the two pulse pairs. Last, a positive auto-correlation contribution always leads to divergent weights. This is summarized in Figure 2.13 where we sketch fixed points (i.e. w_1^∞) of the weight w_1 against the auto-correlation contribution.

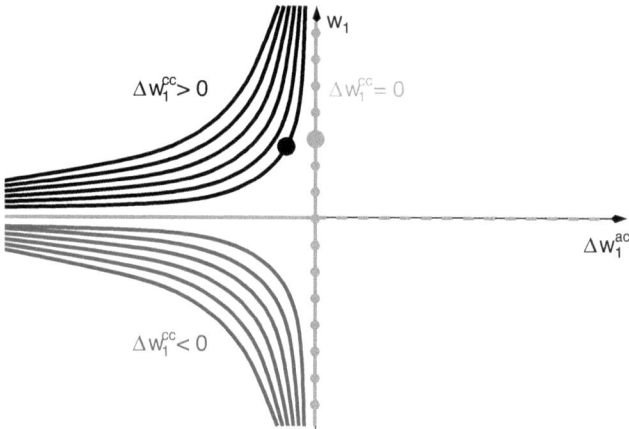

Figure 2.13: Here fixed points of w_1 for different cross-correlation contributions are sketched against the auto-correlation contribution. Fixed points resulting from a positive cross-correlation contribution are plotted in black color, those from negative cross-correlation contribution in darker gray, and without a cross-correlation contribution the fixed points are in light gray. Solid lines indicate stable and dashed lines unstable fixed points.

Relation to spike-timing-dependent plasticity
To what degree are the above discussed models related to temporal sequence learning mechanisms in the brain? For example, we note that the learning curve of ISO learning, ICO learning and some curves of ISO3 learning and VOT plasticity resemble curves measured for spike-timing-dependent plasticity (Markram et al., 1997; Magee and Johnston, 1997; Bi and Poo, 2001). Hence, it is possible to model STDP with such a formalism (Saudargiene et al., 2004; Roberts, 1999). However, there is one problem. Until now we only have looked at neurons with only one plastic synapse, where the other was kept fixed. In real neuronal systems usually more than one synapse is plastic, if not all. Does this make a difference, and if it does, what would thus change? These questions will be covered in the next chapter.

Biophysical aspects of ISO3 and TD learning
The instability of the ISO rule was the reason to design ISO3, which is a form of (differential) Hebbian plasticity using a three-factor learning rule (Miller et al., 1981). Such three-factor rules have recently also been discussed in conjunction with the Dopaminergic system of the brain (Schultz, 1998). Also, since it is a Hebb rule, it is better suited to be matched to our knowledge about LTP and LTD. Furthermore, we found, quite unexpectedly, that for weight stabilization

2.6 DISCUSSION

ISO3 can use one interesting aspect of the behavior of dopamine cells in the substantia nigra and VTA (Schultz et al., 1997). These cells appear to learn to anticipate a reward, whereby the temporal occurrence of their response shifts from t_{x_0} to t_{x_1}. When doing this with our relevance signal in ISO3, learning stops and the weights become essentially stable even without setting $x_0 = 0$ (see chapter 4). Bringing the average TD error δ_r down to zero does require the dopamine responses to take a very specific shape, whereas for stabilizing weights in ISO3 it is enough to roughly adjust the timing. This seems to be better in conjunction with the properties of neuromodulator responses, which do not appear to fulfill high accuracy requirements.

Chapter 3

Many-Plastic-Synapse Systems

Here, we will investigate many plastic synapses, where the change of one synapse influences the plasticity of other synapses. Similar to chapter 2, we will concentrate on differential Hebbian plasticity; however, subsection 3.2.2 is an exception of this rule, where a general solution for linear Hebbian plasticity of many-synapse systems is developed.

Up to now, all rules and figures (up to Figure 2.11) have always shown how the different plasticity rules behave when maximally *one* synapse is plastic. It is, however, important to also know whether the properties we found in the last chapter also hold for multi-synapse systems. There are two different extensions one could think of. In the first extension, we stick to our two signal setup (x_0 and x_1) and extend the number of kernels. This applies if knowledge about the actual timing T is limited. In the previous chapter, we indirectly assumed that we know the temporal difference between the incoming stimuli and if we abandon this assumption, we need to use a set of kernels or eligibility traces for spreading out the earlier stimulus across time to make sure that at least some of these signals can be related to the later occurring x_0 signal. This will be the first situation discussed here.

The other extension is to extend the setup and allow arbitrary input signals x_i, which converge onto all plastic synapses. This will be investigated later on for all linear Hebbian plasticity rules. However, before we extend our system to arbitrary many-synapses, in subsection 3.2.1 we will investigate symmetrical ICO learning (see section 2.4) with two plastic synapses.

3.1 Multiple plastic synapses for a *single* input

The usefulness of all these rules as presented so far remains limited as most of the time the interval T between incoming inputs is not known well enough and might even vary to some degree in a behaving agent. Hence, it is required to use a set of different eligibility traces $h_{1,...,N}$ to make sure that the earlier input is spread out over a sufficiently long time such that the later input (x_0) can be correlated to it. Figure 3.1 A depicts such a kernel bank architecture for the ISO rule; and panel B shows what the signals $u_{1,...,N}$ look like for a set of kernels h.

Interestingly, convergence properties for the ISO rule are theoretically not affected when using a kernel bank. It can be shown that a set of kernels h exists that fulfills certain or-

36 CHAPTER 3 MANY-PLASTIC-SYNAPSE SYSTEMS

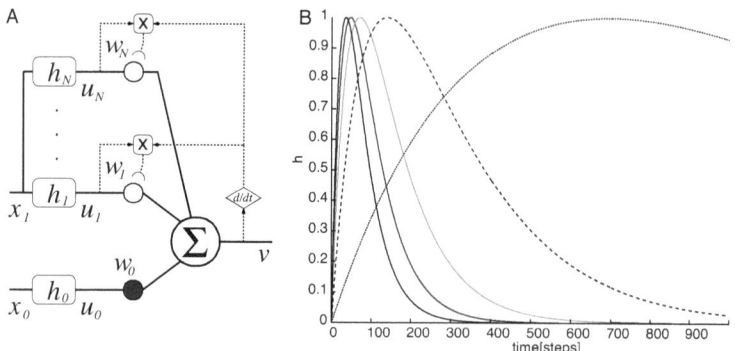

Figure 3.1: *Architecture and example kernels for ISO learning with a bank of kernels. Panel A shows the architecture with different kernels, however, always using the same input x_1. The parameters for the kernels (see equation 1.1) in panel B are $a = 0.001\,\eta$, $b = 0.002\,\eta$, and $\sigma = 0.25$, where the value of η is 20 (black), 15 (darker gray), 10 (light gray), 5 (dashed), and 1 (dotted).*

thogonality criteria, and ISO will then still converge for $x_0 = 0$ (Porr and Wörgötter, 2003a; Porr et al., 2003). The problem is that this is only an existence proof and nothing is currently known of how to actually construct this kernel bank. Hence, when wanting to use ISO, one has to fall back onto heuristic assumptions for the kernel bank. Generally, this leads to the situation that the error sensitivity of ISO can become larger, rendering this rule instable. The properties of ICO and ISO3 are better. The ICO rule is stable per se for $x_0 = 0$, even when using a kernel bank (Porr and Wörgötter, 2006). For ISO3 learning it is possible to eliminate the auto-correlation. This is shown in the following.

Eliminating the auto-correlation of ISO3 learning We will now show that the recursive properties of ISO3 learning using a bank of kernels will self-organize into the constructed case shown in Figure 2.11 A, which was specifically constructed to demonstrate the idea of three-factor learning. This is an extension of Porr and Wörgötter (2007).

A closer look at Figure 2.11 reminds the reader that it was constructed with the first maximum of v exactly at t_0, which is the moment when u_0 sets in. Hence, at the beginning of learning, we get for the left derivative $\dot{v}_{t\to 0_-} = 0$, while the right derivative $\dot{v}_{t\to 0_+} \neq 0$. The idea of this section is to show that the system will self-organize to generically create such a situation and that by this way the auto-correlation term will become zero. The learning rule writes in a more general way having N inputs

3.1 MULTIPLE PLASTIC SYNAPSES FOR A *SINGLE* INPUT

$$\dot{w}_k(t) = \mu\, u_k(t)\, \dot{v}(t)\, R(t) \qquad (3.1)$$

$$\cong \mu \left(u_k(t) \sum_{j=1}^{N} w_j(t)\, \dot{u}_j(t) + u_k(t)\, w_0\, \dot{u}_0(t) \right) R(t) \qquad (3.2)$$

with a non-filtered relevance signal R, which needs to occur at time $T_R = T$: $R(t) = \delta(t - T_R) = \delta(t - T)$. This also applies for $x_0 = \delta(t - T)$, whereas x_1 is set to $x_1 = \delta(t)$. The output generalizes to

$$v(t) = \sum_{j=0}^{N} w_j(t)\, u_j(t). \qquad (3.3)$$

The overall weight change for w_k is in a simplified way (see appendix B and note that the argumentation still holds with an additional factor)

$$\Delta w_k = \mu \int_0^\infty u_k(t)\, \dot{v}(t)\, R(t) dt. \qquad (3.4)$$

This integral is split into a cross- and auto-correlation term so that we get:

$$\Delta w_k = \mu \underbrace{\int_0^\infty u_k(t) \sum_{j=1}^{N} w_j(t)\dot{u}_j(t)\, R(t)\, dt}_{ac_k} + \mu \underbrace{\int_0^\infty u_k(t)\, w_0 \dot{u}_0(t)\, R(t)\, dt}_{cc_k} \qquad (3.5)$$

and is solved by including the delta functions and integrating over them to

$$\Delta w_k = \mu \underbrace{\left(h_k(T) \overbrace{\sum_{j>0} w_j \dot{h}_j(T)}^{\dot{g}_v(T)} \right)}_{ac_k} + \mu \underbrace{h_k(T)\, w_0\, \dot{h}_0(0)}_{cc_k}$$

$$= \mu\, h_k(T)\, \dot{g}_v(T) + \mu\, h_k(T)\, w_0\, \dot{h}_0(0) \qquad (3.6)$$

which means that weight change only occurs at time T.

The second step is to show that at time T the auto-correlation term ac_k remains zero. Since this is a recursive system, we can start with the initial condition $w_k = 0$, $k > 0$, where $ac_k = 0$ (equation 3.6). Hence, at that moment weight development only depends on cc_k. Thus, we need to ask whether weights w_k will from there on develop such that ac_k remains zero, which guarantees stability of the system. Dependency on cc_k renders Δw_k proportional to w_0, $\dot{h}_0(0)$, $h_k(T)$, the plasticity rate, and the number of plasticity experiences W, where only the term $h_k(T)$ changes the distribution of the weights. This means that we replace w_j in the ac_k term in equation 3.6 with $\Lambda\, h_j$ getting

$$g_v(t) = \Lambda \sum_{j=1}^{N} h_j(t) h_j(T) \qquad (3.7)$$

where $\Lambda = W\,\mu\,w_0\,\dot{u}(0)$ accounts for a constant term. Thus, the auto-correlation term will be zero if $\dot{g}_v(T) = 0$. Ultimately, this can only be achieved with an infinite number of kernels so that all possible T are covered, which turns the sum into an integral:

$$g_v(t) = \Lambda \int_0^\infty h_\eta(t) h_\eta(T)\, d\eta \qquad (3.8)$$

where η scales the frequency of the kernels which are defined slightly differently from the previous sections with given rise time a and decay time b

$$h_\eta(t) = \frac{e^{-a\eta t} - e^{-b\eta t}}{\sigma_\eta} = \frac{e^{-a\eta t} - e^{-b\eta t}}{\sqrt{\eta(b-a)}}. \qquad (3.9)$$

We defined the normalization as $\sigma_\eta = \sqrt{\eta(b-a)}$, which will guarantee $ac_k = 0$, as will be shown next. Substituting equation 3.9 into equation 3.8 gives us

$$g_v(t) = \Lambda \int_{\epsilon>0}^\infty \frac{(e^{-a\eta t} - e^{-b\eta t})(e^{-a\eta T} - e^{-b\eta T})}{\eta(b-a)} d\eta \qquad (3.10)$$

where ϵ is infinitely small but non-zero to avoid a singularity in the integral. This amounts to removing the constant component from the frequency distribution of the used kernel bank. The integral equation 3.10 writes as:

$$\begin{aligned}g_v(t) = \Lambda &\left(\int_{\epsilon>0}^\infty \frac{e^{-a\eta t - b\eta T}}{\eta(a-b)} d\eta - \int_{\epsilon>0}^\infty \frac{e^{-a\eta(t+T)}}{\eta(a-b)} d\eta \right.\\ &\left. + \int_{\epsilon>0}^\infty \frac{e^{-a\eta T - b\eta t}}{\eta(a-b)} d\eta - \int_{\epsilon>0}^\infty \frac{e^{-b\eta(t+T)}}{\eta(a-b)} d\eta \right)\end{aligned} \qquad (3.11)$$

These four integrals are essentially of the form[1]

$$E(\xi(t)) = \int_{\epsilon>0}^\infty \frac{e^{-\xi(t)\eta}}{\eta} d\eta \qquad (3.12)$$

and its derivative for $\epsilon \to 0$ is $\dot{E}(\xi(t)) = -\dot{\xi}(t)/\xi(t)$ (see appendix E for a detailed calculation). This results in

$$\dot{g}_v(t) = \frac{\Lambda}{(a-b)} \left(\frac{a}{at+bT} - \frac{a}{a(t+T)} + \frac{b}{aT+bt} - \frac{b}{b(t+T)} \right) \qquad (3.13)$$

and when bringing it in a more compact form with a common denominator, we arrive at the final solution

$$\boxed{\dot{g}_v(t) = \frac{\Lambda T(t-T)(a-b)}{(at+bT)(aT+bt)(t+T)}.} \qquad (3.14)$$

---

[1] These integrals are a special case of the exponential integral $E_n(\xi) = \int_1^\infty e^{-\xi\eta}/\eta^n \cdot d\eta$ with $n = 1$.

3.1 MULTIPLE PLASTIC SYNAPSES FOR A *SINGLE* INPUT

This term becomes zero for $t = T$, which is the desired result rendering the auto-correlation zero at the moment the third-factor signal R is triggered.

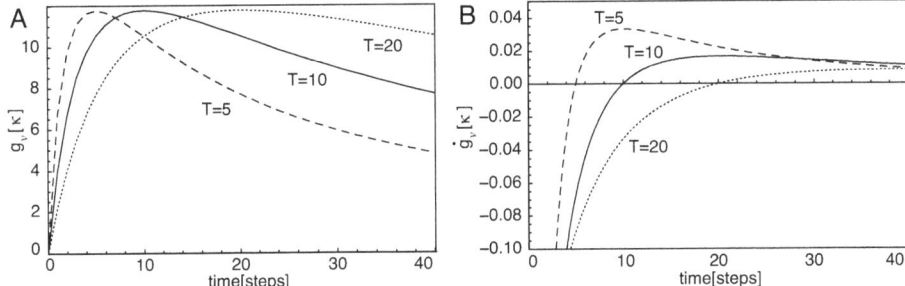

Figure 3.2: *Characteristics of function g_v and its time derivative $\dot{g}_v$ for three different intervals T. The parameters are $\Lambda = 1$, $a = 0.1$, $b = 0.2$ and $T = 5, 10,$ and 20.*

Figure 3.2 A shows a plot of equation 3.13 for different values of T. The choice of a and b is not critical as long as they are not identical. It is clear that the zero crossing is at the desired position $t = T$.

The integral equation 3.10 has no closed form solution, but can be integrated numerically, where the results are shown in Figure 3.2 B. We have chosen $T = 5, 10$ and 20 as the time between x_1 and x_0.

Finally we have to show that at time T the cross-correlation part cc_k is unequal zero, without which no learning would take place. Here, we refer back to the difference of right versus left derivative: $\dot{v}_{t \to 0_-} = 0$ versus $\dot{v}_{t \to 0_+} \neq 0$. Hence cc_k will produce a contribution for $t \to T_+$, which will lead to learning. As a final step, we need to assure that $\dot{g}_v$ is non-divergent around $t = T$. The Taylor expansion around this point

$$\dot{g}_v(t) = \sum_{n=1}^{\infty} \frac{A_n}{T^{n+1}}(t-T)^n \text{ with } A_1 = -\frac{\Lambda(a-b)}{2(a+b)^2} \quad (3.15)$$

shows that for $n = 1$ the function $\dot{g}_v$ follows $1/T^2$, which only results in divergence if $T \to 0$. Note the bigger T is (see different curves in Figure 3.2 A), the more stable is ISO3 learning when not using ideal δ functions.

Hence, we showed that by introducing a bank of kernels which need to follow equation 3.9 with the proposed normalization the auto-correlation contribution stays zero when using a δ-function as relevance signal. However, even if we extend the width of the inputs to a finite value, the auto-correlation stays close to zero which is shown in Figure 3.3. This completes the considerations on many different kernels for the same input.

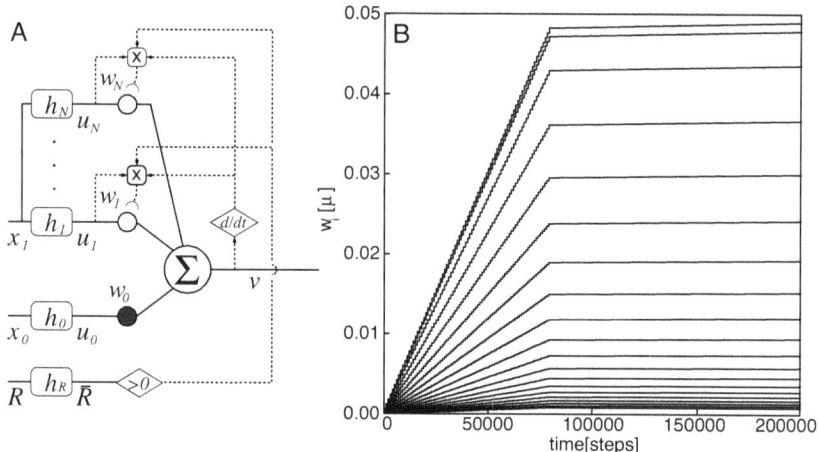

Figure 3.3: *Architecture and weight development of ISO3 learning with a bank of kernels. Panel A shows the architecture where all paths use the same kernel function equation 3.9. In panel B we see the time development of 20 weights w_i, each using the same kernel; however, with different parameters. It was produced by many x_1/x_0 pairs, where the x_0 signal was switched off after time $t = 80000$. Parameters were $w_0 = 1$, $a = 0.01\,i$, $b = 0.02\,i$, $i = 1, 2, \ldots, 20$, $a_R = 4$, $b_R = 8$, and $T = T_R = 20$.*

3.2 Multiple plastic synapses for *many* inputs

Next we drop the single synapse condition and allow all synapses to be plastic. We start with our basic setup having two synapses, which gives us symmetrical architectures. However, we wait with the discussion of symmetrical ISO learning, as the general solution of multi-synapse systems gives us ISO learning as a special case for free. Therefore, symmetrical ICO learning will be the first system to be investigated.

3.2.1 Symmetrical rules: ICO learning

These investigations should also answer whether it is possible to implement learning (LTP) at one synapse and unlearning (LTD) at the other synapse at the same time. In principle this should be possible because one synapse experiences $+T$ while the other experiences $-T$ for any given input pair. Thus, causality is inverted for the two synapses and with the right design one synapse should grow, while the other would shrink.

In Figure 3.4 the isotropic setup for the coupled ICO learning is shown and the plasticity rule is given by:

3.2 MULTIPLE PLASTIC SYNAPSES FOR *MANY* INPUTS

$$\dot{w}_0(t) = u_0(t)\,\dot{u}_1(t)\,w_1(t) \qquad (3.16)$$
$$\dot{w}_1(t) = u_1(t)\,\dot{u}_0(t)\,w_0(t) \qquad (3.17)$$

We solve the weight change analytically to:

$$\Delta w_{0/1} = \mp \frac{b - a\,\text{sign}(T)}{a+b}\,\frac{T}{2\sigma^2}\,h(|T|) \qquad (3.18)$$

which is, except the $\mp$ sign, identical to equation 2.11.

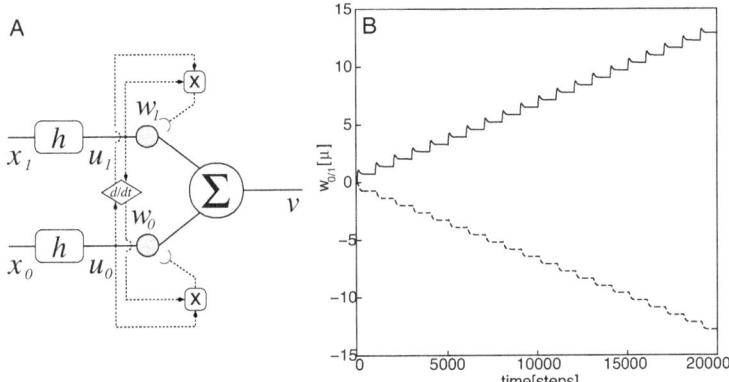

Figure 3.4: *Architecture and weight development of symmetrical ICO learning. Panel A shows the architecture where both paths use the same kernel function. Panel B shows an example time development of many x_1/x_0 pairs, where both weights w_0 (solid) and w_1 (dashed) learn, each starting from a value of 0.1. The time interval between x_1 and x_0 is $T = 60$. Both weights develop in an anti-symmetrical way independent of the time interval used. Parameters were $a = 0.1$, $b = 0.2$, and $\sigma = 0.25$.*

Symmetrical ICO learning (Figure 3.4 A) produces a linear phase-relation, which is not shown here, but Figure 3.4 B shows instead that both weights develop in an anti-symmetrical way. It is interesting that this only holds for the overall weight development. The individual time development is different, as equation 3.31 of subsection 3.2.3 shows.

There is a problem, however; symmetrical ICO learning no longer has one shared control parameter for the weight change, which for symmetrical ISO learning would be the derivative of the output. For symmetrical ICO learning, two totally independent control parameters exist (the derivatives of the inputs). This can possibly lead to problems when wanting to control behavior with such a symmetrized ICO rule.

When looking at ISO learning things are not so simple anymore. Therefore we first develop a complete description, from which we come back to symmetrical ISO learning as a special case.

3.2.2 General many-synapse systems

In this subsection we will follow a different aim as we are not investigating the stability or the convergence of weights. Here we will develop a method to solve such weight development for an arbitrary number of input signals x_i under linear Hebbian plasticity rules. This also means that we generalize in this subsection from differential Hebbian plasticity to all possible linear Hebbian plasticity rules. It is known that in behaving animals sensory inputs are highly non-stationary (Kayser et al., 2003). This generically applies to all systems (animals, machines, robots, etc.) which interact with their environment as their own behavior will lead to continuously changing inputs and, thus, to an ongoing synaptic weight change. The solution provided here may allow for the first time to calculate Hebbian plasticity in such systems without restrictions. Such restrictions are for instance averaging over input signals or neglecting the plasticity of synapses like we have done so far.

The general system is shown in Figure 3.5 on the right side and similar to our previous definitions it consists of N synapses with strength w_i that receive input from neurons i with its continuous values x_i. Each input produces an excitatory post synaptic potential (EPSP), which is modeled by kernel functions h_i (see Figure 1.7). The output of the neuron is, thus:

$$v(t) = \sum_{i=0}^{N} (x_i * h_i)(t) \cdot w_i(t) \tag{3.19}$$

where $(\xi * \eta)(t)$ again describes a convolution. The synapses change according to a general

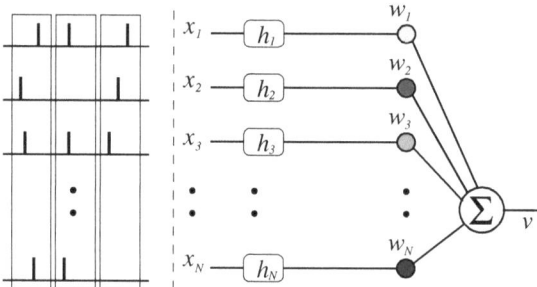

Figure 3.5: *This figure shows our general setup and example input values that are fed to the neuron. Inputs are denoted as x_i, kernel functions as h_i, synaptic strength as w_i and the output of the model neuron as v. The example inputs shown on the left side are spike trains, although any arbitrary continuous function can serve as an input. Note that the plasticity pathway is not shown.*

formalized Hebbian plasticity rule

$$\dot{w}_i(t) = \mu \, F[x_i * h_i](t) \, G[v](t) \tag{3.20}$$

3.2 MULTIPLE PLASTIC SYNAPSES FOR *MANY* INPUTS

where $F[\cdot]$ and $G[\cdot]$ are linear functionals.

We already know conventional Hebbian plasticity with $F = G = \mathbf{1}$ (where $\mathbf{1}$ is the identity) and differential Hebbian plasticity with $F = \mathbf{1}$ and $G = \frac{d}{dt}$.

To avoid that weight changes will follow spurious random correlations we also assume that plasticity is a slow process, where inputs change much faster than weights, with $\frac{dw_i}{w_i} \ll \frac{d(x_i * h_i)}{x_i * h_i}$, $\mu \to 0$. This simplifies equation 3.20 and we neglect all temporal derivatives of w_i on the right hand side:

$$\dot{w}_i(t) = \mu \, F[x_i * h_i](t) \sum_{j=0}^{N} w_j(t) \, G[x_j * h_i](t) \tag{3.21}$$

where we used $G[\sum \xi_i] = \sum G[\xi_i]$ as $G[\cdot]$ is linear.

If we take w_i as the i-th component of a vector $\boldsymbol{w}$, we write

$$\dot{\boldsymbol{w}}(t) = \mu \, \boldsymbol{A}(t) \, \boldsymbol{w}(t) \tag{3.22}$$

with $A_{ij}(t) = F[x_i * h_i](t) \, G[x_j * h_i](t)$ or in matrix form

$$\boldsymbol{A}(t) = F[(\boldsymbol{x} * h)(t)] \cdot G[(\overline{\boldsymbol{x}} * h)(t)] \tag{3.23}$$

$$= \begin{pmatrix} F[(x_0 * h)(t)] \, G[(x_0 * h)(t)] & \cdots & F[(x_0 * h)(t)] \, G[(x_N * h)(t)] \\ \vdots & \ddots & \vdots \\ F[(x_N * h)(t)] \, G[(x_0 * h)(t)] & \cdots & F[(x_N * h)(t)] \, G[(x_N * h)(t)] \end{pmatrix}$$

where $\overline{\boldsymbol{\xi}}$ denotes the transposition of matrix $\boldsymbol{\xi}$.

The solution of equation 3.22 is not trivial as the matrix $\boldsymbol{A}(t)$ is also a function of time. This problem is often found in quantum mechanics, and the main problem is that matrices usually do not commute. However, there exists a solution which includes an infinite series, called the Magnus series (see Magnus (1954) for more details), with

$$\boldsymbol{w}(t) = \exp \boldsymbol{\Omega}(t) \cdot \boldsymbol{w}_0 \tag{3.24}$$

where $\boldsymbol{w}_0$ is the synaptic strength before plasticity, and $\boldsymbol{\Omega}(t)$ is the solution of following equation

$$\dot{\boldsymbol{\Omega}}(t) = \left\{ \mu \boldsymbol{A}(t), \frac{\boldsymbol{\Omega}(t)}{1 - \exp(-\boldsymbol{\Omega}(t))} \right\} = \sum_{n=0}^{\infty} \beta_n \left\{ \mu \boldsymbol{A}(t), \boldsymbol{\Omega}^n(t) \right\}. \tag{3.25}$$

Here the braces $\{\eta, \xi^n\} = [\cdots[[\eta, \xi], \xi]\cdots\xi]$ are nested commutators $[\eta, \xi] = \eta\xi - \xi\eta$ and β_n are the coefficients of the Taylor expansion of $\frac{\Omega}{1-\exp(-\Omega)}$ around $\boldsymbol{\Omega} = 0$. equation 3.25 is solved through integration by iteration to the Magnus series:

$$\boldsymbol{\Omega}(t) = \mu \boldsymbol{\mathfrak{A}}(t) + \frac{\mu^2}{2} \int_0^t [\boldsymbol{A}(z_1), \boldsymbol{\mathfrak{A}}(z_1)] \, dz_1$$
$$+ \frac{\mu^3}{4} \int_0^t \left[\boldsymbol{A}(z_1), \int_0^{z_1} [\boldsymbol{A}(z_2), \boldsymbol{\mathfrak{A}}(z_2)] \, dz_2 \right] dz_1 + \frac{\mu^3}{12} \int_0^t [[\boldsymbol{A}(t), \boldsymbol{\mathfrak{A}}(z_1)], \boldsymbol{\mathfrak{A}}(z_1)] \, dz_1$$
$$+ o(\mu^4) \tag{3.26}$$

with $\boldsymbol{\mathfrak{A}}(t) = \int_0^t \boldsymbol{A}(z)dz$. Thus, equation 3.24 combined with equation 3.26 gives us analytically the time development of all weights connected to a neuron under Hebbian plasticity in the limit of small plasticity rates μ. With this, we are principally able to calculate the synaptic strengths of N synapses without simulations, given N different spike trains, membrane potentials, or firing rates.

Next we transform the solution into a computable form and provide error estimates. As the commutators in the infinite series in equation 3.26 are generally non-zero we truncate the series and neglect iterations above degree (k). We write the truncated solution as:

$$\boldsymbol{w}_{(k)}(t) = \exp \boldsymbol{\Omega}_{(k)}(t) \cdot \boldsymbol{w}_0 \qquad (3.27)$$

For two synapses, this is solved directly in the next subsection; most often, however, equation 3.27 needs to be calculated by expanding the exponential function. We denote this approximation with a prime, i.e. (k')

$$\boldsymbol{w}_{(k')}(t) = \left(\boldsymbol{I} + \sum_{p=2, q=1}^{p \cdot q \leq k} \left(\boldsymbol{\Omega}_{(p)}(t) \right)^q \right) \cdot \boldsymbol{w}_0 = \boldsymbol{\mathfrak{B}}_{(k')}(t) \cdot \boldsymbol{w}_0 \qquad (3.28)$$

where $\boldsymbol{I}$ is the identity matrix and $\boldsymbol{\mathfrak{B}}_{(k')}(t)$ the transformation of order (k) from the initial synaptic strength $\boldsymbol{w}_0$ to the synaptic strength at time t. This solution is computable for arbitrary input patterns. Notice that in the limit $k \to \infty$ the approximation (equation 3.27) transforms into the general solution (equation 3.24).

Now as we know the complete analytical solution of equation 3.22 we investigate the approximations and their errors in order to judge their usefulness for further considerations. We will switch to spikes as the inputs to the system and assume that all $h_i = h$ are equal. As earlier, spikes are modeled as delta functions $\delta(t - t_i)$ for spike times t_i and the convolution is simplified to a temporal shift in the kernel function h: $h(t - t_i)$. This leads to $A_{ij}(t) = F[h](t - t_i) G[h](t - t_j)$ for elements of $\boldsymbol{A}(t)$ where t_i and t_j are the spike timings of neuron x_i and x_j respectively. We will use our standard kernel functions (equation 1.1).

The different approximation errors are exemplified in Figure 3.6. For this, we are using a single spike pair at two synapses, for which we calculate the final synaptic strength $\hat{\boldsymbol{w}} = \lim_{z \to \infty} \boldsymbol{w}(z)$ (equation 3.24). This has been performed for differential Hebbian plasticity, but we point out that the error is identical for Hebbian plasticity as the order of the error is not affected by the actual (linear) Hebbian rule used. This is because the only source of the error is the plasticity rate μ, which is independent from the choices of F and G. For this setup weight changes are computed in three ways: without any approximations, yielding $\hat{\boldsymbol{w}}$ (equation 3.24 and equation 3.26); using the truncated solution only, yielding $\hat{\boldsymbol{w}}_{(k)}$ (equation 3.27); and using the truncated solution while also expanding the exponential function, yielding $\hat{\boldsymbol{w}}_{(k')}$ (equation 3.28). Thus, we use $\hat{\boldsymbol{w}}$ and compare it to approximations $\hat{\boldsymbol{w}}_{(\cdot)}$, calculating the error as: $\Delta_{(\cdot)} = \left| \hat{\boldsymbol{w}}_{(\cdot)} - \hat{\boldsymbol{w}} \right|$. This is plotted in Figure 3.6 against the plasticity rate μ for different approximations on a log-log scale. As approximations (k) and (k') become very similar for $k > 2$, only four curves are

3.2 MULTIPLE PLASTIC SYNAPSES FOR *MANY* INPUTS

shown. We observe that the behavior of the difference-error $\Delta_{(\cdot)}$ follows the order of the approximation used. The error for the linear expansion approximation ($k = 2'$, equation 3.28) is slightly higher than that of its corresponding truncation approximation ($k = 2$, equation 3.27). However, using a plasticity rate of $\mu = 0.001$ already results only in a difference-error value of 10^{-8} as compared to 10^{-2} when using $\mu = 1$. Therefore, in most applications one can use even the simplest possible linear approximation ($k = 2'$) to calculate the change in synaptic strength.

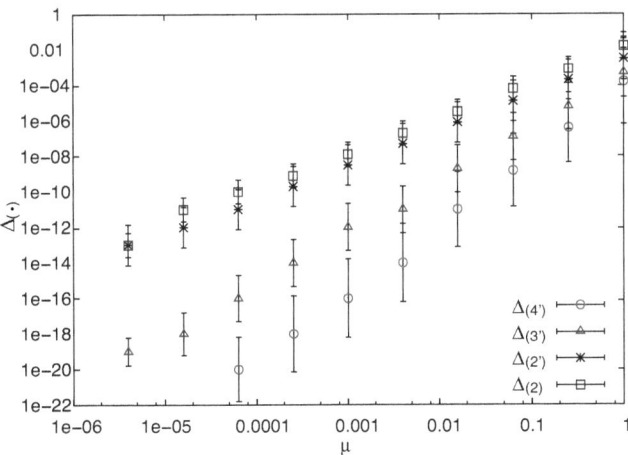

Figure 3.6: *Here we show the degree of consistency between our general solution and the proposed approximations. To this end, we plot the difference $\Delta_{(\cdot)}$ between the approximation and the exact solution of equation 3.22 for one input spike pair against the plasticity rate μ on a log-log scale. A kernel function h with $a = 0.1$, $b = 0.2$, $\sigma = 0.25$, and $\max_t h(t) = 1$ is used. The temporal difference $T = t_1 - t_0$ between the two input spikes was varied over the length of the used kernel functions (here between 1 and 100 steps), and error bars representing the standard deviation are given.*

As this calculation has been based on two spikes at two synapses only, we need to ask how the error develops when using N synapses and complex spike trains. For this we first consider spike trains (see Figure 3.5 left), which are grouped 'vertically' into groups with each input firing at most once. Kernels of spikes within a group will overlap, but we assume that grouping is possible such that adjacent groups are spaced with a temporal distance sufficient to prevent overlap between kernel responses of temporally adjacent groups. Thus we calculate $\boldsymbol{w}$ in the same way as above, leading to: $\hat{\boldsymbol{\mathfrak{B}}}_{(k')} = \lim_{z \to \infty} \boldsymbol{\mathfrak{B}}_{(k')}(z)$ in equation 3.28. When using such a temporal tiling, $\hat{\boldsymbol{\mathfrak{B}}}_{(k')}$ depends only on the spike timing matrix $\boldsymbol{T}$ with elements $T_{ij} = t_j - t_i$, and we get the synaptic strength after H groups by calculating the product over all groups m:

$$\boxed{\boldsymbol{w}_{M,(k')} = \prod_{m=1}^{H} \hat{\mathfrak{B}}_{(k')}(\boldsymbol{T}_m) \cdot \boldsymbol{w}_0.} \quad (3.29)$$

Physiologically such a grouping decomposition can be performed for so-called non-bursting neurons, which, for example, constitute the majority of cortical cells. The solution (equation 3.29) is easy to compute. As a product of matrices results in a summation of matrix elements, the error does not increase exponentially but only linearly in M. Due to this it follows that even after 10000 spikes the error is still of an order of only 10^{-4} given any of the approximations in the example above (see Figure 3.6). Thus, the easily computable group decomposition suggested by equation 3.29 will yield results accurate enough even for long, non-bursting spike trains.

Finally we estimate how the error behaves when kernels overlap. This mainly happens during bursts of spikes with temporarily high spiking frequencies which in general are rare events. However, using the solution which assumes independent temporal intervals (equation 3.29) instead of the time-continuous calculation (equation 3.28) only includes an additional error of order ($k = 2$) due to the linearity of the kernel functions h. The error after matrix multiplication (equation 3.29) results in the square of the lowest term of the Magnus series (equation 3.26).

In this subsection we elaborated analytical solutions with different degrees of approximation. Finally we will briefly compare the algorithmic complexity of the required analytical versus the numerical calculations. In general, for the analytical solutions we need to perform a matrix multiplication for every moment in time at which we want to analyse the weights. This multiplication costs $2 \cdot N \cdot N$ steps with N being the number of inputs. The numerical calculation is linear in N (see appendix F) and, thus, costs $9 \cdot N + n$ calculations where we have $n = 1$ for plain Hebbian plasticity and $n = 2$ for differential Hebbian plasticity as this demands an additional subtraction. Thus, with more than six inputs the numerics are preferable as it only costs 54 computations per weight calculation compared to 72 for the analytics. This, however, is only the naive assessment of the algorithmic complexity. Realistically, we can always benefit from the fact that analytics are to be calculated "per spike" whereas numerics require calculations "per sampling step". This difference can be dramatic especially for many synapses with sparse spike trains. Numerical calculations require fine sampling to avoid numerical errors (see appendix C) and the step size Δt needs to be very small (e.g. 10^{-3}). If we liked to compute the synaptic changes after 10000 spike groups (as defined for equation 3.29) with an average firing rate of 2 Hz, similar to cortical cells, having a sampling rate of 10^3 Hz, we would need to perform the complete numerical integration procedure for $10000/2\,\text{Hz} \cdot 10^3\,\text{Hz} = 5 \cdot 10^6$ times. By contrast, for the analytics we would only need 10000 complete calculations. In general, if S is the number of sampling steps that can be skipped between time t and time $t + S \cdot \Delta t$, the analytical solution is advantageous as soon as we exceed $S = N$. This shows that the analytical solution is in almost all cases preferable as compared to numerics.

3.2 MULTIPLE PLASTIC SYNAPSES FOR *MANY* INPUTS

3.2.3 Symmetrical rules: ISO learning

Next we look in more detail at symmetrical homosynaptical differential Hebbian plasticity, i.e. ISO learning with two plastic synapses (see Figure 3.7), which is analytically fully solvable. We have also based the error analysis provided in Figure 3.6 on these calculations. For this case the matrix $\mathfrak{B}(t)$ results in

$$\mathfrak{B}(t) = \begin{pmatrix} 1 + \frac{\mu}{2} h^2(t) & \mu\, \nu_{T,-1}(t) \\ \mu\, \nu_{T,+1}(t) & 1 + \frac{\mu}{2} h^2(t - T) \end{pmatrix} \qquad (3.30)$$

where $\nu_{T,-1}(t) = \int_0^t h(z)\, \dot{h}(z - T)\, dz$ and $\nu_{T,+1}(t) = \int_0^t h(z - T)\, \dot{h}(z)\, dz$. Here we use two input neurons ($N = 2$), which received a spike at $t = 0$ and at $t = T$ respectively.

Using the kernel function h (equation 1.1), we analytically integrate the secondary diagonal entries of equation 3.30 which are:

$$\begin{aligned}
\nu_{T,\eta}(t) = &\frac{\Theta(t - T)\,\Theta(t)}{2\,(a + b)\,\sigma^2} (\eta\, \text{sign}(T)\, \sigma\, (a - b)\, h(|T|) \\
&- 2 e^{-t(a+b)} (a e^{aT} + b e^{bT}) \\
&+ (a + b)(e^{-a(2t - T)} + e^{-b(2t - T)})).
\end{aligned} \qquad (3.31)$$

which is identical to equation 2.12 for $\eta = -1$.

In the limit of t to infinity matrix $\mathfrak{B}(t)$ changes into $\hat{\mathfrak{B}}$ and so do the secondary diagonal elements (compare with equations 2.11, 2.24, and 3.18)

$$\hat{\nu}_{T,\eta} = \lim_{t \to \infty} \nu_T(t) = \eta\, \text{sign}(T)\, \frac{a - b}{2\,(a + b)\,\sigma} h(|T|). \qquad (3.32)$$

Furthermore, we find that $\hat{\nu}_T = \hat{\nu}_{T,+1} = -\hat{\nu}_{T,-1}$. For the considered kernel function $\hat{\nu}_T$ is positive definite as a is smaller than b. Therefore $\hat{\mathfrak{A}}$ results in

$$\hat{\mathfrak{A}} = \lim_{t \to \infty} \mathfrak{A}(t) = \begin{pmatrix} 0 & \hat{\nu}_T \\ -\hat{\nu}_T & 0 \end{pmatrix} = \nu_T \begin{pmatrix} 0 & 1 \\ -1 & 0 \end{pmatrix}. \qquad (3.33)$$

The diagonal elements become zero as the chosen kernel function decays to zero in the limit to infinity.

As the square is $\hat{\mathfrak{A}}^2 = -\hat{\nu}_T^2\, I$, we calculate the exponential solution equations 3.27 for an error of order ($k = 2$). The exponential function is then:

$$\begin{aligned}
\hat{\mathfrak{B}}_{(2)} &= \exp \mu\hat{\mathfrak{A}} = \sum_{n=0}^{\infty} \frac{1}{n!} (\mu \hat{\mathfrak{A}})^n = \sum_{n=0}^{\infty} \frac{(-1)^n}{(2n)!} (\mu\, \hat{\nu}_T)^{2n}\, I + \sum_{n=0}^{\infty} \frac{(-1)^n}{(2n+1)!} (\mu\, \hat{\nu}_T)^{2n+1}\, J \quad (3.34) \\
&= \cos(\mu\, \hat{\nu}_T)\, I + \sin(\mu\, \hat{\nu}_T)\, J = \begin{pmatrix} \cos(\mu\, \hat{\nu}_T) & \sin(\mu\, \hat{\nu}_T) \\ -\sin(\mu\, \hat{\nu}_T) & \cos(\mu\, \hat{\nu}_T) \end{pmatrix}
\end{aligned}$$

where $\boldsymbol{J} = \begin{pmatrix} 0 & 1 \\ -1 & 0 \end{pmatrix}$. This results in

$$\hat{\boldsymbol{w}}_{(2)} = \hat{\boldsymbol{\mathfrak{B}}}_{(2)} \cdot \boldsymbol{w}_0 = \begin{pmatrix} \cos(\mu\,\hat{\nu}_T) & \sin(\mu\,\hat{\nu}_T) \\ -\sin(\mu\,\hat{\nu}_T) & \cos(\mu\,\hat{\nu}_T) \end{pmatrix} \boldsymbol{w}_0. \quad (3.35)$$

Both equation 3.33 and equation 3.35, were used to calculate the difference $\Delta_{(\cdot)}$ for different values of T in Figure 3.6.

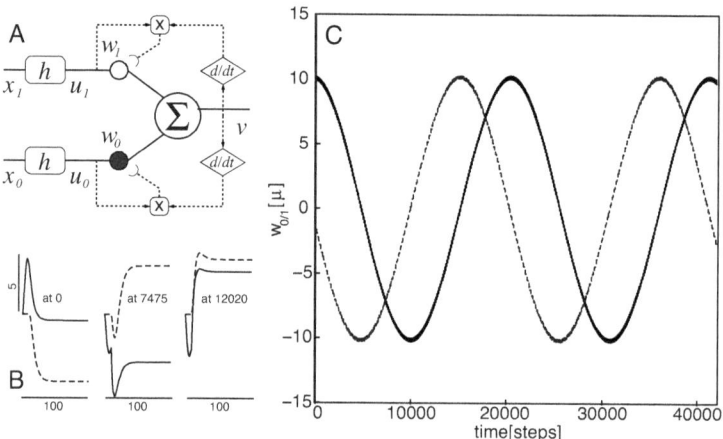

Figure 3.7: *Architecture and weight development of symmetrical ISO learning. Panel A shows the architecture, where both paths use the same kernel function. Panel C shows an example time development of many x_1/x_0 pairs, where both weights w_0 (solid) and w_1 (dashed) learn starting from a value of 10 and −1, respectively. The time interval between x_1 and x_0 was $T = 10$. In Panel B close-ups of the time development at different times are shown. Parameters were $a = 0.1$, $b = 0.2$, $\sigma = 0.25$, and $\mu = 0.05$.*

Having found the analytical solution, we can now investigate the weight development of symmetrical ISO learning (see Figure 3.7 A). This is plotted in Figure 3.7 C, where we find that both weights oscillate around zero. In Panel B, close-ups of the development are shown. The second close-up shows clearly how the second weight interacts and thus changes the development of the first weight. It is interesting that the amplitude of the individual time development (panel B) is plasticity rate dependent and, by contrast, the amplitude of the overall development (panel C) is not. However, as equation 3.35 gives us a coupled differential equation which is easy to solve, we determine the frequency ω of the oscillation to $\omega = \sin(\mu\,\hat{\nu}_T)$, which is again plasticity rate dependent.

3.3 Discussion

In this chapter we investigated the plasticity of multiple synapses, however, in two different systems. In the first, we only had *one* input that splits into many pathways, where all have different kernels. By contrast the second system had *many* inputs, each also influencing a different weight. For both systems there exist solutions, however, the solution for the second setup is more complex. This is because besides the interaction of synapses that change all the time comes the interaction of varying input patterns.

Single input

For different rules, namely ISO, ICO and ISO3 learning, proofs exists that the auto-correlation contribution can be eliminated (Porr and Wörgötter, 2003a, 2006, 2007). Additionally, they have now been successfully tested in a variety of different applications (Porr and Wörgötter, 2003b; Kolodziejski et al., 2006, 2007; Manoonpong et al., 2007), and even chains of learning neurons can be constructed in a convergent way (Kulvicius et al., 2007). All these applications show that the extension to multiple kernels, hence to multiple plastic synapses, is important if the temporal shift between the two inputs $x_{0/1}$ was not known.

Many inputs

Real neurons often display rich, non-stationary firing patterns, by which all synaptic weights will be affected. The same is true for neurons in artificial neural networks, especially when embedded in closed-loop (acting, behaving) systems. The solutions existing so far which describe Hebbian learning, on the other hand, restrict the temporal dynamics of the system or limit plasticity to a subset of synapses (see chapter 2). With the solution presented here, we can calculate weight changes time without these restrictions for the first. This is a valuable step forward in our understanding of synaptic dynamics in different networks. Specifically, we have presented the time-continuous solution for the synaptic change of general Hebbian plasticity (equation 3.24 and equation 3.26), its approximation for general spiking or continuous inputs (equations 3.27 and 3.28) as well as a specific solution for non-bursting spike trains (equation 3.29). Of practical importance is the fact that the error of the computable approximations (equations 3.27, 3.28, 3.29) remains small even for long spike trains.

The temporal development of multi-synapse systems and the conditions of stability are still not well understood. Some convergence conditions have been found (see for example Hopfield (1982); Miller and MacKay (1994); van Rossum et al. (2000); Roberts (2000); Kempter et al. (2001); Burkitt et al. (2007)); however, in general the synaptic strengths of such networks will diverge or oscillate. This is undesired, because network stability is important for the formation of (e.g.) stable memories or receptive fields. Using the time-continuous solution for linear Hebbian plasticity described here, could serve as a starting point to better understand mechanisms, structures and conditions for which stable network configurations will emerge. The rich dynamics, which govern many closed-loop adaptive (network based) physical systems can, thus, now be better understood and predicted, which might have substantial future influence for the guided design of network controlled systems.

Chapter 4

The Relation of Differential Hebbian Plasticity to Reinforcement Learning

In the last chapter, we described the mathematical properties of differential Hebbian plasticity or learning. We put this kind of plasticity in the context of real neurons, trying to point to the similarities between synaptic plasticity and differential Hebbian plasticity and in particular the properties that relate to spike-timing-dependent plasticity. Further on, when talking about learning, we gave examples about the relation of differential Hebbian learning and Pavlovian (or classical) conditioning. One should ask whether it is not possible to extend this relation to operant (or instrumental) conditioning in a straightforward manner. There, the action of the learner influences the stimuli the system receives and, in turn, the stimuli by means of corresponding weights decide about the next action to take. We have already discussed the first aspect in the Pavlovian context of closed-loop systems (see section 1.2). Now we will discuss operant conditioning. With reinforcement learning (see Sutton and Barto (1998) for an overview), the change of values is guided by rewards or punishments (negative rewards), and the actions that lead to the reward are reinforced. Hence, this kind of learning is the method of choice to model operant conditioning.

Given a stimulus or rather a state and its weight learned by differential Hebbian plasticity, we can not evaluate by means of the current weight strength what comes next. That the weight is positive just tells us that there is positive correlation with another state, thus with some probability there will be another state. To judge whether a given state predicts something "good" is only possible if an evaluative process adjust a certain value, for instance the corresponding weight. However, correlation-based learning of which also (differential) Hebbian plasticity is part, is only able to handle non-evaluative changes of weights as neither an explicit (supervised learning) nor an implicit (reinforcement learning) error signal is used. Indeed it belongs to the class of unsupervised learning rules.

In order to realize reinforcement learning in a biological more realistic way, we need to embed the learning rule into a network structure. For instance, we presented an implementation of temporal difference learning, which belongs to the class of reinforcement learning algorithms, in section 2.3. There the δ error (see equation 2.21) is an entity that can not be computed directly at the neuron but needs to be calculated elsewhere in the network. On the other hand, weight changes in (differential) Hebbian plasticity follow biophysically realistic mechanisms, namely the correlation of pre- and post-synaptic activation. Now the question arises whether it is possible to emulate the first, reinforcement learning, with the latter, differential Hebbian plasticity. In order to emulate reinforcement learning, we will concentrate on a widely used algorithm, i.e. temporal difference learning, and give a more technical (i.e. machine learning) introduction in the following.

Emulating reinforcement learning by temporal difference Learning Reinforcement learning maximizes the rewards $r(x)$ an agent will receive in the future when following a policy π traveling along states x. The return R is defined as the sum of the future rewards: $R(x_i) = \sum_k \gamma^k r(x_{i+k+1})$, where future rewards are discounted by a factor $0 < \gamma < 1$ i.e. a reward n time steps in the future is only worth γ^n. One central goal of RL is to determine the values $V(x)$ for each state given by the average expected return $E^\pi\{R\}$, which can be obtained when following policy π. Many algorithms exist to determine the values, almost all of which rely on the temporal difference (TD) learning rule (equation 4.1) (Sutton, 1988).

Every time the agent encounters a state x_i, it updates the value $V(x_i)$ with the discounted value $V(x_{i+1})$ and the reward $r(x_{i+1})$ of the next state associated with the consecutive state x_{i+1}:

$$V(x_i) \to V(x_i) + \alpha \left[r(x_{i+1}) + \gamma V(x_{i+1}) - V(x_i) \right] \tag{4.1}$$

where α is the learning rate. This rule is similar to equation 2.20, however, with two differences. First, weight w and output v are represented by the same value V and secondly, as equation 4.1 does not represent neuronal activity, there is no dependence on the pre-synaptic input u anymore. This rule is the general TD($\lambda = 0$) rule, short TD(0), as it only evaluates adjacent states. For values of $\lambda \neq 0$ more of the recently visited states are used for value-function update. TD(0) is by far the most influential RL learning rule as it is the simplest way to assure optimality of learning (Dayan and Sejnowski, 1994; Sutton and Barto, 1998).

Whether such a learning algorithm does what it is supposed to do, i.e. weights develop and converge in the right way, is an important question and convergence proofs exist for many algorithms (see Hertz et al. (1991) for an overview). It is even possible that different algorithms fulfill the same task although the method used and the solution found are completely different. Alternatively, it is possible that, although the methods are not conform, the solution is. Two methods are asymptotically (after convergence) equivalent if at least in the mean the solutions are identical.

Recently there have been several contributions towards finding the equivalence between spike-timing-dependent plasticity and the here discussed temporal difference learning. Izhikevich (2007); Roberts et al. (2009); Florian (2007); Potjans et al. (2009) presented specific solutions, which we will discuss in more detail in section 4.3. Thus, there is more and more evidence emerging that Hebbian plasticity (such as STDP) and reinforcement learning can be brought together under a more unifying framework. Such an equivalence would have substantial influence both on our understanding of network learning and the biological mechanisms of reinforcement learning as these two types of learning could be interchanged under certain conditions.

Thus, the goal of this chapter is to prove that the temporal difference (TD) learning rule (Sutton, 1988), is asymptotically equivalent to differential Hebbian plasticity under certain rather general conditions, like having a negative auto-correlation contribution. This can be achieved either by using a third factor (in a global way - subsection 4.2.1 or in a local way - subsection 4.2.2) or by using different time scales for the plasticity and the output pathway (subsection 4.2.3). We will also show that equivalence holds over wide parameter ranges.

Biophysical considerations about how such a third factor might be implemented in real neural tissue are of secondary importance for this thesis. At this stage we are concerned with a formal proof only. Some biophysical aspects have already been treated in section 2.3 and will be extended in section 4.3, though.

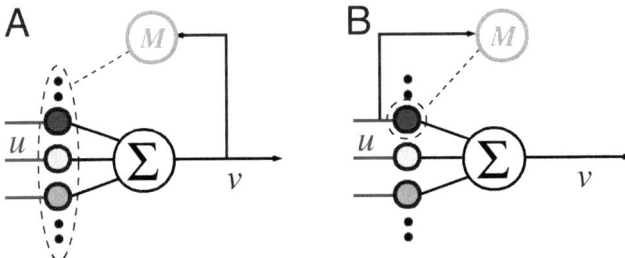

Figure 4.1: *Here the difference between a global and a local third factor M is exemplified. Panel A shows the global third factor, where the output activity v drives the third factor M which in turn influences all synaptic connections. In contrast, panel B depicts a local third factor M triggered by the input u and only influencing a single synaptic connection.*

Concerning the third factor, there are two different ways to use it and this also modifies the influence of the third factor on the weights. These two ideas are depicted in Figure 4.1. The first method to use the third factor is through the output activity of a neuron. Here the third factor will globally influence all synaptic connections that converge onto this neuron, thus calling it a *global third factor* (Figure 4.1 A). On the other hand, it is also possible to use the input activity as a trigger for the third factor. With this method only the synaptic connection, whose input triggered the third factor, will be affected. Because now the third factor acts locally, it is called a *local third factor* (Figure 4.1 B).

In the following we will start with the more general global third factor and continue with the local third factor by pointing out differences in our analytics. We will find that the local third factor has advantages over the global factor with respect to convergence and computations. The last part of our analysis section shows a possibility to achieve equivalence without a third factor. We will, however, see that a shorter time scale at the output pathway acts similarly to a third factor.

The next section discusses the aspects of the setup that are fundamentally important to achieve the equivalence between differential Hebbian plasticity and temporal difference learning. Having this in mind, we show the equivalence for the three above-mentioned modifications of the basic differential Hebbian plasticity rule. To achieve this we solve the resulting differential equation and use these to show the asymptotic equivalence in a straightforward manner. The solution of the differential equation gives us constraints which we will investigate for generally applicable signal shapes. A simulated network will then show some practical aspects before we finish with a technical discussion of the given modifications. A general discussion is given at the end of this chapter.

4.1 General setup

In the introduction to this chapter we saw that in temporal difference learning both a TD value V and a reward r is assigned to a discrete state x. This is shown in Figure 4.2 A, where we see three states $(x_1 - x_3)$ having non-zero TD values V and one state x_r having a non-zero reward r. For simplification we do not show the zero-valued rewards and the TD value at x_r. As temporal difference learning is an algorithm which uses discrete time and space, the agent travels per time unit from one state to the next. On the other hand, we have Hebbian-like

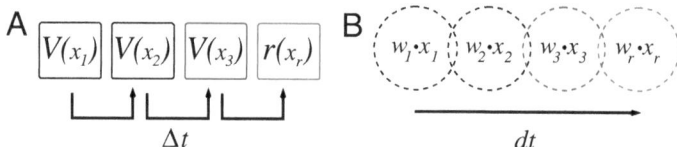

Figure 4.2: *Typical state structure used for temporal difference learning (A) and the here proposed setup (B). Temporal difference learning usually uses discrete time and space (indicated by the broken arrows and Δt). By contrast, the Hebbian plasticity operates in continuous time and space (indicated by the continuous arrow and dt). Both panels show four states having either a value V or a reward r. The corresponding variables for the here proposed setup are plastic weights w_i and fixed weights w_r, respectively. The discrete states correspond to continuous regions, where the corresponding input x_i is active.*

plasticity which uses the correlation of an inputs x and an output v to change a synaptic weight w. In panel B of Figure 4.2 we depict in continuous space the area with circles which corresponds to an input x_i. Whenever the agent enters a circle, the corresponding input will be active and the corresponding weight changes continuously in time. Thus, if we want to emulate temporal difference learning with Hebbian-like plasticity, we need to relate states to inputs and values to weights. This is indicated in Figure 4.3. Along these lines rewards are represented

4.1 GENERAL SETUP

by *rewarded* states which have fixed non-adaptive weights, which is their reward value. That these weights are fixed and non-zero represents the fact that rewards are usually associated with unconditioned stimuli, which will lead to strong, insuppressible responses.

Continuous versus discrete space and time As Hebbian-like plasticity uses continuous time, in the following we have to integrate over the whole time where the signal u of state x is significantly unequal zero. This then results in the new weight value of the corresponding state x. The temporal difference between two states is thus achieved through the cross-correlation contribution between temporal adjacent states and their signals. Therefore it makes sense to rearrange in the following the states into a temporal and not a spatial order, i.e. state x_{i+1} comes after state x_i although these states might be located far apart.

Here, we introduce another constraint concerning the rewarded states x_R which we define as terminating states. Therefore a temporal state sequence always ends with a rewarding state x_R. The drawback of this constraint will be discussed at the end of this section.

We already implied that there is a relation between states and signals and similar to the last chapters we convolve states $x(t)$ with a kernel function h, which leads to our signal $u(t) = \int_0^\infty x(z) \, h(t-z) \, dz$. We define the kernel functions to be identical for all states. Note that in subsection 4.2.3 we differentiate between the plasticity and the output pathway (compare to VOT plasticity in section 2.1). This will change the parameters a and b of the kernel function h (see equation 1.1), which will be used for the output pathway. We will indicate this with an index v. As we are using only states that are either on or off during a long enough visiting duration S, the input functions $u(t)$ essentially do not differ between states. Therefore we will use $u_i(t)$ (with index i) having a particular state in mind and $u(t)$ (without index i) when pointing to functional development. If we assume, guided by biophysics, that our kernels are stereotypic with a rising phase of length P_E and a falling phase of length P_F (compare Figure 1.7 where $P_F = P_E$), long enough means that the signal was able to reach the plateau before falling again, hence $S > P_E$.

Hence, in general the overall change of the weight w_i after integrating over the significant non-zero values of u_i (i.e. simplified over the visiting duration S of $x_{i-1}(t)$, $x_i(t)$ and $x_{i+1}(t)$) results in $\Delta w_i =: \Delta_i = \Delta_i^{cc-} + \Delta_i^{ac+} + \Delta_i^{ac-} + \Delta_i^{cc+}$. Note that we defined $\Delta_i^{ac\pm}$ as $\Delta w_i^{ac\pm} \cdot w_i$.

Specifications Without loss of generality we are going to analyze the change of weight w_i when considering three consecutive signals u_{i-1}, u_i and u_{i+1}. There are two time windows which are important for the auto-correlation as well as for the cross-correlation contribution. For the latter the first window opens with the beginning of the falling phase of the preceding signal u_{i-1} and, because of the negative slope of u_{i-1}, we define it as $cc-$. The second window opens with the start of the rising phase of the subsequent signal u_{i+1} and is defined as $cc+$. Similarly, for the auto-correlation contribution the first window opens with the rising phase of the according signal u_i and is defined as $ac+$ and the second goes with the falling phase of u_i, thus named $ac-$. As mentioned in the introductory paragraphs of this chapter, a possible third factor M might open more and/or different time windows. However, we will define M to essentially be active "around" the windows discussed above assuring that states correlate with temporally neighboring states.

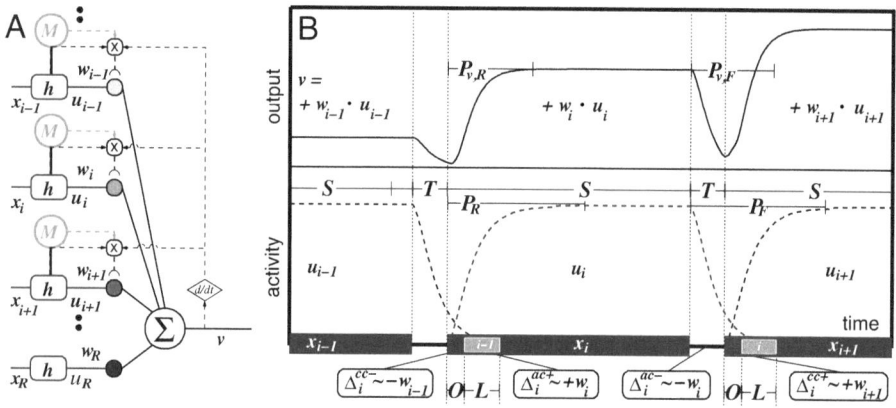

Figure 4.3: *The general setup is shown in panel A, and an arbitrary signal structure in panel B. (A) Three temporal adjacent states and the rewarded state converge on the neuron which learns according to a differential Hebbian plasticity rule. Plasticity at synapse w_i is influenced by pre-synaptic, by post-synaptic, and by possible modulatory activity. The states x will be active in an increasing order. (B) The lower part shows the states x which have an activity duration of length S. We assume that the duration for the transition between two states is T. Above the signals u and the output v are depicted. Here we additionally indicated the duration of the rising (P_E and $P_{v,E}$) and the falling phase (P_F and $P_{v,F}$) of the signals and the output, respectively. Signals u are given by $u(t) = \int_0^S (e^{-a(t-z)} - e^{-b(t-z)})\,dz$. For state x_i the weight change contributions of the auto-correlation $\Delta^{ac\pm}$ and cross-correlation $\Delta^{cc\pm}$ are indicated.*

We already defined S as the length of each state x. Furthermore, we define the period between the end of a state $x_i(t)$ and the beginning of the next state $x_{i+1}(t)$ as T, where $T < 0$ represents overlapping states. We define O as the onset time of a modulatory factor M and L as its duration. Two (S and T) of these four constant parameters (S, T, O, and L) are displayed in detail in Figure 4.3 B. The remaining parameters are depicted in the particular setup figures (Figure 4.4 and Figure 4.8) in the following sections.

Importance of a negative auto-correlation In order to emulate temporal difference learning with a Hebb-like plasticity rule using this general setup, a negative auto-correlation contribution is a necessary condition. This is the central idea behind the following proof which arises from the analysis made in chapter 2. There, we found that a negative auto-correlation leads to convergent non-zero weights if the input activity is correlated with successive non-vanishing inputs.

4.2 General analysis

Using the general setup defined in the last section, we analyze the underlying differential weight change equation, the equivalence and for which parameter ranges the weights converge. This general analysis is followed by specific implementations, namely differential Hebbian plasticity with a global third factor, with a local third factor, and with different time scales for plasticity and output. For this we further solve and investigate the equations we received from the general analysis.

Analysis of the differential equation Plain differential Hebbian plasticity will not suffice for our purposes as the auto-correlation contribution is per se equal to zero, and thus not negative (see section 2.1). Hence, we need to use differential Hebbian plasticity with a modification which will be a third factor (compare to ISO3, section 2.5) or different time scales for the plasticity and the output pathway (compare to VOT plasticity, last paragraph of section 2.1). This leads in a general form to

$$\dot{w}_k(t) = \tilde{\alpha}\, u_k(t)\, \dot{v}(t)\, M(t) \quad (4.2) \qquad v(t) = \sum_l w_l(t) u_{v,l}(t) \quad (4.3)$$

where (similar to chapter 2) $u_k(t)$ is the considered pre-synaptic signal and $v(t)$ the post-synaptic activity of a model neuron with weights $w_k(t)$. The index v at the pre-synaptic activity indicates that a different kernel or rather different parameters are being used for the output as compared to plasticity. We will assume in the following that our modulatory signal M is either on or off, i.e. 1 or 0. Thus it is represented by a step function.

By means of a learning rate $\tilde{\alpha}$ we can set the ratio between the weight change over the weight $\dot{w}/w$ to be significantly smaller than the state change of the state value $\dot{u}/u$. Hence, we assume a quasi-static process ($\frac{\dot{w}_i}{w_i} \ll \frac{\dot{u}_i}{u_i}$, $\tilde{\alpha} \to 0$) with all the consequences that are discussed in appendix B.

For the following analysis we need to substitute equation 4.3 in equation 4.2 and solve this differential equation which consists of a homogeneous and an inhomogeneous part:

$$\dot{w}_i(t) = \tilde{\alpha}\, M(t)\, u_i(t) \frac{d}{dt}[w_i(t)\, u_{v,i}(t)] + \tilde{\alpha}\, M(t)\, u_i(t) \frac{d}{dt}[\sum_{j \neq i} w_j(t)\, u_{v,j}(t)]. \quad (4.4)$$

We will use this equation which consists of equations 4.2 and 4.3 for the following practical emulations of temporal difference learning; however, always adapted to the particular conditions made by each of the methods. Thus, it makes sense to list and compare these equations by pointing to the differences. In each of the sections covering a particular method we will refer to table 4.1.

The integration boundaries are defined by either the intrinsic properties of the signal shape (rising/falling phase duration P_E and P_F respectively - see Figure 4.3) or the modulatory factor M. This is generalized by means of a bounded temporal path π. The first summand leads to the homogeneous solution which we already defined in chapter 2 as auto-correlation $w^{ac}(t)$.

Method	Rule $\dot{w}_i(t)$	Output $v(t)$	Combined $\dot{w}_i(t)$
Global	$\tilde{\alpha}\, u_i(t)\, \dot{v}(t)\, M(t)$ (4.5)	$\sum_j w_j(t) u_j(t)$ (4.6)	$\tilde{\alpha}\, M(t)\, u_i(t)\, \dfrac{d}{dt}[w_i(t)\, u_i(t)]$ $+\tilde{\alpha}\, M(t)\, u_i(t)\, \dfrac{d}{dt}[\sum_{j\neq i} w_j(t)\, u_j(t)]$ (4.7)
Local	$\tilde{\alpha}\, u_i(t)\, \dot{v}(t)\, M_i(t)$ (4.8)	$\sum_j w_j(t) u_j(t)$ (4.9)	$\tilde{\alpha}\, M_i(t)\, u_i(t)\, \dfrac{d}{dt}[w_i(t)\, u_i(t)]$ $+\tilde{\alpha}\, M_i(t)\, u_i(t)\, \dfrac{d}{dt}[\sum_{j\neq i} w_j(t)\, u_j(t)]$ (4.10)
VOT	$\tilde{\alpha}\, u_i(t)\, \dot{v}(t)$ (4.11)	$\sum_j w_j(t) u_{v,j}(t)$ (4.12)	$\tilde{\alpha}\, u_i(t)\, \dfrac{d}{dt}[w_i(t)\, u_{v,i}(t)]$ $+\tilde{\alpha}\, u_i(t)\, \dfrac{d}{dt}[\sum_{j\neq i} w_j(t)\, u_{v,j}(t)]$ (4.13)

Table 4.1: *Overview over the essential equations used to emulate temporal difference learning. In particular, 'Global' is short for the global third factor, 'Local' for the local third factor and 'VOT' for the implementation with different time scales. Concerning the rule (second column) the difference between the global and the local third factor is at the index of the modulatory factor M. The index is needed because the local third factor modulates only the corresponding weight, thus the weight with the same index. By contrast, as the modulatory factor in the VOT rule does not exist, we can mathematically set it to 1 or even simpler, just neglect it. The output (third column) is identical for the local and the global third factor. However, as the VOT rule needs different time scales for plasticity and output, we change the kernel parameters (a and b) of the output convolution, which is indicated by the subindex v. In the last column we substituted the output in the rule which results in the equation we have to consider when implementing the rule.*

The second summand(s) on the other hand will lead to the inhomogeneous solution and this was defined as cross-correlation $w^{cc}(t)$. Together we have $w(t) = w^{ac}(t) + w^{cc}(t)$.

Using the argument of a quasi-static process, we neglect the derivatives of w on the right hand side of equation 4.4. The solution of the auto-correlation $w_i^{ac}(t)$ is then in general (see appendix D):

$$w_i^{ac}(t) = w_i(t_0) \exp\left(\tilde{\alpha} \int_\pi u_i(z)\, \dot{u}_{v,i}(z)\, dz\right) \qquad (4.14)$$

4.2 GENERAL ANALYSIS

where t_0 is defined as the lower bound of π. The overall weight change is therefore:

$$\Delta_i^{ac} = w_i \left(\exp\left(\tilde{\alpha} \int_\pi u_i(z)\, \dot{u}_{v,i}(z)\, dz \right) - 1 \right) \tag{4.15}$$

Because of the quasi-static process ($\tilde{\alpha} \to 0$), we expand the exponential function to the first order:

$$\Delta_i^{ac} = -\tilde{\alpha}\, w_i\, \kappa + o(\tilde{\alpha}^2) := \tilde{\alpha}\, w_i \int_\pi u_i(z)\, \dot{u}_{v,i}(z)\, dz + o(\tilde{\alpha}^2). \tag{4.16}$$

where we have defined κ to be positive:

$$\begin{aligned}\kappa(\pi) &= -\int_\pi u_i(z)\, \dot{u}_{v,i}(z)\, dz \\ &= -\left(\kappa^+(\pi_{ac+}) + \kappa^-(\pi_{ac-})\right).\end{aligned} \tag{4.17}$$

Note that κ corresponds to a *negative* auto-correlation contribution: $-\Delta w^{ac}$. Here we have also left out the index i as all states are identical and split κ into κ^+ and κ^- representing the first temporally-bounded path π_{ac+} (positive slope of signal u - see Figure 4.3 B) and the second temporally-bounded path (negative slope of signal u - see Figure 4.3 B). To this end, as mentioned before, we also have to split Δ_i^{ac} into $\Delta_i^{ac+} = \tilde{\alpha}\, w_i\, \kappa^+$ and $\Delta_i^{ac-} = \tilde{\alpha}\, w_i\, \kappa^-$.

Next we investigate the cross-correlation $w^{cc}(t)$ again under the assumption of a quasi-static process. This leads us to:

$$w_i^{cc}(t) = w_i(t_0) + \tilde{\alpha}\, w_{i-1} \int_{\pi_{cc-}} u_i(z)\dot{u}_{v,i-1}(z)\, dz + \tilde{\alpha}\, w_{i+1} \int_{\pi_{cc+}} u_i(z)\dot{u}_{v,i+1}(z)\, dz \tag{4.18}$$

where we split the temporally-bounded path π into π_{cc-} for the preceding state and π_{cc+} for the consecutive state. Furthermore, as all signals are identical, we can shift between signals by $t = S + T$. In detail this is $u_{i-1}(t) = u_i(t + S + T)$ and $u_{i+1}(t) = u_i(t - S - T)$. The overall weight change can then be split into Δ_i^{cc-} and Δ_i^{cc+}:

$$\begin{aligned}\Delta_i^{cc} &= \tilde{\alpha}\, w_{i-1} \int_{\pi_{cc-}} u_i(z)\dot{u}_{v,i}(z+S+T)\, dz \\ &\quad + \tilde{\alpha}\, w_{i+1} \int_{\pi_{cc+}} u_i(z)\dot{u}_{v,i}(z-S-T)\, dz \end{aligned} \tag{4.19}$$

$$=: \underbrace{\tilde{\alpha}\, w_{i-1}\, (-\tau^-)}_{\Delta_i^{cc-}} + \underbrace{\tilde{\alpha}\, w_{i+1}\, \tau^+}_{\Delta_i^{cc+}} \tag{4.20}$$

Here we defined $\tau^\pm$ as being positive and independent of i:

$$\tau^-(\pi_{cc-}) = -\int_{\pi_{cc-}} u(z)\, \dot{u}_v(z+S+T)\, dz \tag{4.21}$$

$$\tau^+(\pi_{cc+}) = \int_{\pi_{cc+}-S-T} u(z+S+T)\, \dot{u}_v(z)\, dz. \tag{4.22}$$

CHAPTER 4 RELATION TO REINFORCEMENT LEARNING

Both $\tau^{\pm}$ and κ depend on the actually used signal shapes u, u_v and the temporally-bounded path given by either the modulatory factor or again the signal shapes.

Analysis of the equivalence Without restrictions, we can now limit the discussion to the situation in Figure 4.3 A, where we have a state transition from x_{i-1} via x_i to x_{i+1}. The state x_{i+1} is either an arbitrary state or the reward. Thus, differential Hebbian plasticity will influence the synaptic connections w_i of states x_i which directly project onto neuron v. In Figure 4.3 B beneath the signals we indicate the different contributions (Δ values) to the overall weight change defined above.

Here we consider the weight change of w_i in more detail. This results from the transition between x_{i-1} and x_i, from the visiting state x_i itself and from the transition between x_i and x_{i+1}. The short learning period at the beginning of the signal u_i will cause a negative weight change Δ_i^{cc-} because of the correlation between the negative derivative of u_{i-1} and the positive value of u_i. Additionally, there is a weight change Δ_i^{ac+} caused by the signal itself. Due to the positive slope of the signal u_i at the beginning of the state, the auto-correlation contribution will be positive. The next learning interval occurs when the state x_i has been left and the signal u_i already decays. This negative slope results in a negative auto-correlation contribution. The fourth contribution yields a positive weight change Δ_i^{cc+} because the positive derivative of the next state signal u_{i+1} correlates with the positive value of signal u_i of state x_i.

Such a sequence exists also when the next state transition occurs, yielding contributions for the Δ_{i+1} values. During the first trial (where all weights are zero) only the cross-correlation Δ_j^{cc+} of state x_j which comes before the rewarding state yields a contribution due to the finding of the reward.

In general the weight after a single trial is the sum of the old weight w_i and the four Δ_i values:

$$w_i \rightarrow w_i + \Delta_i^{ac-} + \Delta_i^{ac+} + \Delta_i^{cc-} + \Delta_i^{cc+} \tag{4.23}$$

Using equations 4.16 and 4.20 we can reformulate equation 4.23 into

$$w_i \rightarrow w_i + \tilde{\alpha}\left(\kappa^+ + \kappa^-\right)w_i - \tilde{\alpha}\,\tau^-\,w_{i-1} + \tilde{\alpha}\,\tau^+\,w_{i+1} \tag{4.24}$$

Substituting $\kappa = -(\kappa^- + \kappa^+)$, $\alpha = \tilde{\alpha} \cdot \kappa$ and $\gamma^{\pm} = \tau^{\pm}/\kappa$, we get

$$w_i \rightarrow w_i - \alpha\,\kappa\,w_i - \alpha\,\gamma^-\,w_{i-1} + \alpha\,\gamma^+\,w_{i+1} \tag{4.25}$$

The convergence to $w_i = \gamma\,w_{i+1}$ is a property of these kind of equations (see appendix G). According to equation G.9 of appendix G, γ needs to be replaced by λ^{-1}, in addition to $\varepsilon_1 = \tau^+/\kappa = \gamma^+$ and $\varepsilon_2 = \tau^-/\kappa = \gamma^-$, provided the values of κ and $\tau^{\pm}$ are strictly positive. These conditions will be discussed in the next paragraph. This gives us:

$$\frac{1}{\gamma} = \frac{1}{2\gamma^-} + \sqrt{\frac{1}{(2\gamma^-)^2} + \frac{\gamma^+}{\gamma^-}}, \tag{4.26}$$

4.2 GENERAL ANALYSIS

and our weight development can be simplified to:

$$w_i \rightarrow w_i - \alpha\, w_i + \alpha\, \gamma\, w_{i+1} \qquad (4.27)$$

At this point we can make the transition from weights w_i (differential Hebbian plasticity) to states $V(x_i)$ (temporal difference learning). Additionally, we note that sequences only terminate at $i+1$, thus this index will capture the reward state x_R and its value $r(x_{i+1})$, while this is not the case for all other indices (see end of this section for a detailed discussion of rewards at non-terminal states). Consequently this gives us an equation almost identical to equation 4.1:

$$V(x_i) \rightarrow V(x_i) + \alpha\, [\gamma\, r(x_{i+1}) + \gamma\, V(x_{i+1}) - V(x_i)] \qquad (4.28)$$

where one small difference arises as in equation 4.28 the reward is scaled by γ. However, this has no influence as numerical reward values are arbitrary. Thus, if learning follows a Hebbian-like plasticity rule with a negative auto-correlation contribution ($\kappa > 0$), weights will converge to the optimal estimated TD values.

Analysis of the convergence Next we will take a closer look at κ (equation 4.17) and $\tau^\pm$ (equations 4.22 and 4.21) as well as, resulting from this, γ (equation 4.26). For guaranteed convergence, these values are constrained by two conditions (see appendix G), $\tau^\pm \geq 0$ and $\kappa \geq 0$, where $\kappa = 0$ is allowed only in case of $\tau^\pm = 0$. A non-positive value of κ would lead to divergent weights w and negative values of $\tau^\pm$ to oscillating weight pairs (w_i, w_{i+1}). However, even if fulfilled, these conditions will not always lead to meaningful weight developments. In particular, $\tau^\pm$ values of 0 leave all weights at their initial weight value, and discount factors which are represented by γ values exceeding 1 are usually not considered in reinforcement learning (Sutton and Barto, 1998). Thus it makes sense to introduce more rigorous conditions and demand that $0 < \gamma \leq 1$ and $\kappa > 0$. In the following sections we will discuss particular implementations which allows to determine these values. In order to discriminate between the different methods, we will indicate the κ, τ and γ values with indices which are: G (global third factor), L (local third factor), and T (different time scales).

Technical discussion Here we will cover basic technical constraints and discuss these with respect to the derivation of the equivalence presented above.

Quasi-static process, $\alpha \ll 1$. In the derivation the assumption of a quasi-static process has been used three times:

1. First, we used this assumption for solving the differential equation (equation 4.4) of the weight change. More precisely we neglected the derivative of the weight on the right side of equation 4.4. If we also considered this term, we would get an inverse square root function $1/\sqrt{1 - \alpha\, \nu}$ instead of the exponential function $e^{\alpha\,\frac{1}{2}\nu}$ (see appendix B for a more

detailed derivation of the differential equation and equation 4.14) we assumed during our proof; the parameter ν is defined here as:

$$\nu := \int_\pi u_i(z)\,\dot{u}_{v,i}(z)\,dz = -\kappa(\pi). \tag{4.29}$$

The inverse square root function, however, has similar properties and expands equally (compare appendix B) with respect to the first order around small values of α compared to the exponential function, justifying constraint $\alpha \ll 1$ here.

2. Second, we truncated the expansion of equation 4.15 after the first order, which is also only allowed for $\alpha \ll 1$. Would the necessary condition $\kappa > 0$ be affected if we had not truncated the expansion? Considering ν as defined in equations 4.29 and 4.17, we observe that a positive value of κ corresponds to a negative value of ν and will lead to a negative weight change of Δ_i^{ac}. Hence, given a negative value of ν (which is a necessary condition for $\kappa > 0$ if taking only the first order terms of the expansion into account), this leads directly to a negative weight change of Δ_i^{ac} in equation 4.15. This is due to the properties of the exponential function ($e^\eta - 1 < 0 \;\; \forall \;\; \eta < 1$) or of the inverse square root function ($1/\sqrt{1-\eta} - 1 < 0 \;\; \forall \;\; \eta < 1$), and, as a consequence, constraining to $\alpha \ll 1$ is allowed here as well.

3. Third, we neglected, because of $\alpha \ll 1$, the variability of the homogeneous solution (w^{ac}, see equation 4.14) in order to calculate the inhomogeneous solution (w^{cc}, see equation 4.18 and appendix B) of the weight w. However, taking the variability into consideration will not affect the linearity with respect to w. This is because equation 4.20 can be directly split into $\tau^\pm$ and w, and the additional homogeneous solution will only change the integral (equations 4.21 and 4.22) which leads to $\tau^\pm$ as the solution does not rely on w.

<u>Reward only at the end of a sequence.</u> In most physiological experiments (Schultz et al., 1992; Montague et al., 1996; Morris et al., 2006) the reward is given at the end of the stimulus sequence. Our assumption that the reward state is a terminating state and is therefore only at the end of the learning sequence, conforms, thus, to this paradigm. However, for TD in general we can not assume that the reward is only provided at the end. Differential Hebbian plasticity will then lead to a slightly different solution compared to TD learning. This solution has already been discussed in a another context (Dayan, 2002). Specifically, the difference in our case is the final result for the state value after convergence for states that provide a reward: We get $V(x_i) \to \gamma V(x_{i+1}) + r(x_{i+1}) - r(x_i)$ compared to TD learning: $V(x_i) \to \gamma V(x_{i+1}) + r(x_{i+1}))$. It would be interesting to assess with physiological and or behavioral experiments which of the two equations does more closely represent experimental reality. To do so one has to guarantee that the reward given at the end is worth the costs that the animal incurred until reaching it (Hassani et al., 2001).

<u>Finite number of states.</u> If we just consider a finite number of states without periodic boundary conditions and assume that always the same state neuron x_0 starts the whole sequence, the corresponding weight will not converge to $w_0 = \gamma w_1$ but to $w_0 = \gamma^+ w_1$ due to the missing

4.2 GENERAL ANALYSIS

w_{-1} weight (see equations 4.27 and G.5). However, in this case the gradual order of the weight values is only disturbed if the γ^+ value is larger than 1.

Non-Markov. A system is Markovian if future states the system will visit only depend on the state the system is currently in, i.e. the system is independent of past states. In the presented algorithm, each TD value arises from the interaction of *three* states. Hence for a considered state it is not negligible, from where it had been reached. Thus, strictly speaking, the algorithm is history dependent and violates the Markov-property required for TD learning (Sutton and Barto, 1998). This has no substantial consequence, as the standard trick for non-Markovian systems, which often are also encountered in conventional RL problems, can be applied here in the same way: A network can easily be designed where states in a higher network layer are being concatenated into new larger states, which now obey the Markov property. In many practical applications, this is not even required as the value-gradient field will build up towards the reward, regardless of the non-Markovian algorithm presented here. An implementation of TD learning on a linear network (see last paragraph of subsection 4.2.1) using our algorithm behaves in this way. By contrast, we will show that the local third factor fulfills the Markovian property.

4.2.1 Global third factor

As described in the introduction, the third factor can be modeled in two different ways (see Figure 4.1). This has not only an influence on the mathematical properties of the proof but also on the conditions of convergence. Although the local third factor both is more straightforward in the mathematics and exhibits advantageous computational properties, we will start with the more general global third factor.

The specific setup is depicted in Figure 4.4 A, where the third factor triggered by the output v influences all synaptic connections uniformly. The corresponding signal structure is shown in Figure 4.4 B. The third factor M gets always triggered at the beginning of a state x and is switch on after an onset time O. After time L it is switched off again. Thus, the third factor M defines the boundaries of π. With this we are now able to determine and calculate κ, $\tau^{\pm}$ and finally the discount factor γ. We will indicate these values calculated in this subsection with an index G (global).

Analysis of the differential equation The underlying equations 4.5, 4.6 and 4.7 can be found in table 4.1 on page 58. They are identical to equation 2.25 of section 2.5 covering ISO3 learning. Only the third factor was replaced by M (modulatory) to avoid collisions between the reward r (or rather the return R) and the relevance signal R.

The boundaries of the temporal path π are $t = O$ and $t = O + L$ for $ac+$ and $t = S + T + O$ and $t = S + T + O + L$ for $ac-$. As we are now using the same kernels for the plasticity

CHAPTER 4 RELATION TO REINFORCEMENT LEARNING

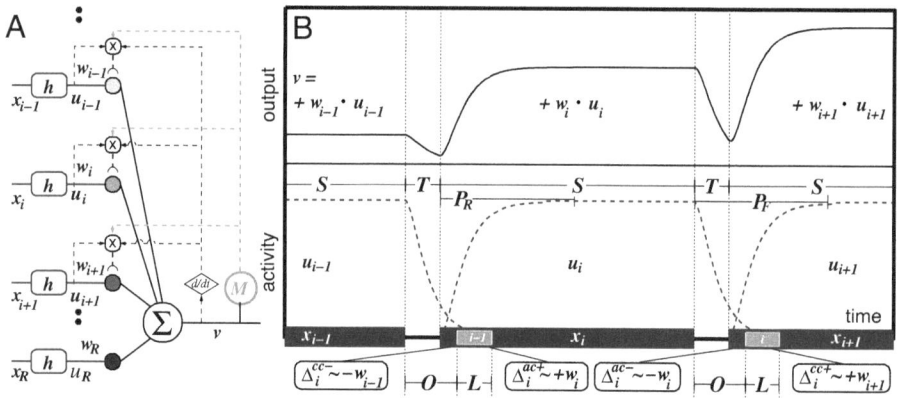

Figure 4.4: *The setup is shown in panel A and the signal structure in panel B. (A) Three states and the rewarded state converge on the neuron which learns according to equation 4.5. The modulatory factor M will influence plasticity at all synapses w_i. The states x will be active with increasing indices. (B) The lower part shows the states x_i which have a duration of length S. We assume that the duration for the transition between two states is T. Above the output v and the signals u are depicted. We additionally indicated the duration of the rising (P_E) and the falling phase (P_F) of the signals. Note that the duration for the output and the plasticity pathway are equal. Here u is given by $u(t) = \int_0^S (e^{-a(t-z)} - e^{-b(t-z)})\,dz$. The third factor M is released for the duration L after an onset time of O and is also shown in the lower part. For state x_i the weight change contributions of the auto-correlation $\Delta^{ac\pm}$ and cross-correlation $\Delta^{cc\pm}$ are indicated.*

and the output pathway, the integral of equation 4.17 simplifies, according to equation D.5 in appendix D, to $\frac{1}{2}u^2(t)$. We now have to include the boundaries, which results in

$$\begin{aligned}\kappa_G(S,T,O,L) =& \frac{1}{2}\left(u^2(O) - u^2(O+L)\right) \\ &+ \frac{1}{2}\left(u^2(S+T+O) - u^2(S+T+O+L)\right) \\ =& -\left(\kappa_G^+(O,L) + \kappa_G^-(S,T,O,L)\right).\end{aligned} \quad (4.30)$$

Thus the Δ_i^{ac}-function split into $\Delta_i^{ac+} = \tilde{\alpha}\,\kappa_G^+\,w_i$ and $\Delta_i^{ac-} = \tilde{\alpha}\,\kappa_G^-\,w_i$.

For the cross-correlation contribution we include the same boundaries discussed above into equations 4.22 and 4.21, which leads to

$$\tau_G^-(S,T,O,L) = -\int_O^{O+L} u(z)\,\dot{u}(z+S+T)\,dz \quad (4.31)$$

$$\tau_G^+(S,T,O,L) = \int_O^{O+L} u(z+T+S)\,\dot{u}(z)\,dz \quad (4.32)$$

4.2 GENERAL ANALYSIS

Here the Δ_i^{cc}-function split into $\Delta_i^{cc-} = -\tilde{\alpha}\,\tau_G^-\,w_i$ and $\Delta_i^{cc+} = \tilde{\alpha}\,\tau_G^+\,w_i$.

Both $\tau_G^\pm$ and $\kappa_G^\pm$ depend on the actually used signal shape $u(t)$ and the values for the parameters S, T, O and L.

Analysis of the equivalence After having calculated $\tau_G^\pm$ and $\kappa_G^\pm$ this paragraph is not different from section 4.2 except that we have to add the index G to κ, τ and γ. Thus, if learning follows this global third factor differential Hebbian rule, weights will converge to the optimal estimated TD values. This proves that, under some conditions for the signal shape and the parameters S, L, O and T (which influence whether $\kappa_G > 0$ and $\tau_G^\pm > 0$), TD(0) and the here proposed global three factor differential Hebbian plasticity are indeed asymptotically equivalent.

Analysis of the convergence Now we will cover the conditions for the signal shape and the parameters (S, T, O and L), which will lead to the requirement that γ_G should be between zero and 1 ($0 < \gamma_G \leq 1$) and that κ_G should be strictly positive ($\kappa_G > 0$).

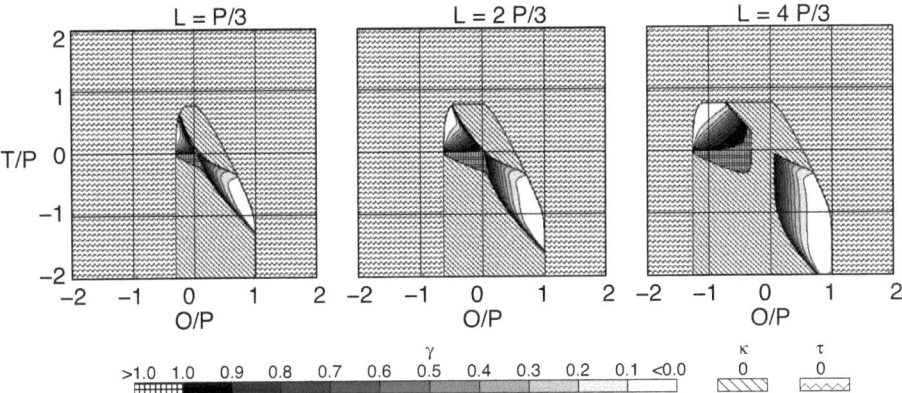

Figure 4.5: *Shown are γ_G values dependent on the ratio O/P and T/P for different values of L/P (1/3, 2/3 and 4/3). Here P is the length of the rising as well as the falling phase. The shape of the signal u is identical to the one used in Figure 4.3 and Figure 4.4 and is given by $u(t) = \int_0^S (e^{-a(t-z)} - e^{-b(t-z)})\,dz$ with parameters $a = 0.006$ and $b = 0.066$. The individual figures are subdivided into a striped area where the weights will diverge ($\kappa_G \leq 0$, see equation 4.30), a zig-zag area where no overlap between both signals and the third factor exists and into a checked area that consists of γ_G values which, however, are beyond a meaningful range ($\gamma_G > 1$). The detailed gray shading represent γ_G values ($0 < \gamma_G \leq 1$) for which convergence is fulfilled.*

As already discussed in the introductory section of this chapter, the theoretical considerations need to be guided by biophysics. Hence, we will discuss neuronally plausible signals that can arise at a synapse. This limits u to functions that possess only one maximum and divide the signal into a rising and a falling phase with length P_E and P_F respectively.

One quite general possibility for the shape of the signal u is the function used in Figure 4.3 and Figure 4.4 for which we investigate the area of convergence. We have three parameters to be varied as we do not have to consider the parameter S if we take this value to be large compared to $|T|$, O or L. For this, Figure 4.5 shows the γ_G value in 3 different panels. In each panel we varied the parameters O and T from minus to plus $2P$ where $P = P_E = P_F$ is the time the signal u needs to reach the maximum (or fall to zero). In each of the panels, we plot γ_G values for a particular fraction of L/P.

A gray shading displays in detail the γ_G values, for which the condition of convergence is fulfilled, whereas checked represents those areas, for which we receive $\gamma_G > 1$. The zig-zag area indicates parameter configurations for which no overlap exists between two consecutive signals and the third factor ($\tau_G = 0$), and for the striped regions, κ_G is smaller than zero.

If the L value is greater than $P - O - T$, the area of convergence does not depend on L anymore as the third factor then reaches a plateau as well as covers the whole falling phase of the signal u. On the contrary, if the L value reaches the rising phase of the consecutive state, the area of convergence decreases again (not shown).

For positive O values there exist γ_G values which are independent of (negative) T values. Hence, if states overlap ($T < 0$), the γ value is invariant with respect to the degree of overlap. This is an important aspect as value function approximation methods often use overlapping kernels to represent features. In a biological context, this corresponds to overlapping receptive fields providing the input to the system. We find that in these cases γ_G remains unaffected by the degree of (receptive field) overlap, which in general is different for any two input units.

To extend these considerations to more general but smooth signal shapes, we Taylor expand both the rising and the falling phases to the second order. With these constraints γ_G can be calculated analytically (see appendix H) and is then plotted in Figure 4.6 with respect to O and T for nine different input functions shown in the lower right. In the upper left panel, the ratio between the duration of the third factor and P was set to $1/3$, in the upper right to $2/3$, and the lower left to $4/3$. Analogous to the exponential function, the area of convergence increases with increasing L values. Figure 4.6 reveals that the biophysically most realistic shape (bottom right) also has the largest convergence range.

The analytical calculations in appendix H are also used to extract information about the areas in which the algorithm diverges ($\kappa \leq 0$) or in which the weights of systems do not change at all ($\tau = 0$). This allows us to put these areas together to depict regions where γ_G is either convergent or divergent. Figure H.7 can then be compared with the results of Figure 4.6. It shows that both figures match each other as we have used the same derivations in the appendix as for Figure 4.6. However, even if you use more general kernel functions, such as the exponential function used for Figure 4.5, both figures still match quite well, especially in the regions where the system diverges and where it stays constant.

In summary the different figures (4.5, 4.6 and H.7) show clearly that the area of convergence changes only gradually and the area as such increases with increasing duration of the third fac-

4.2 GENERAL ANALYSIS

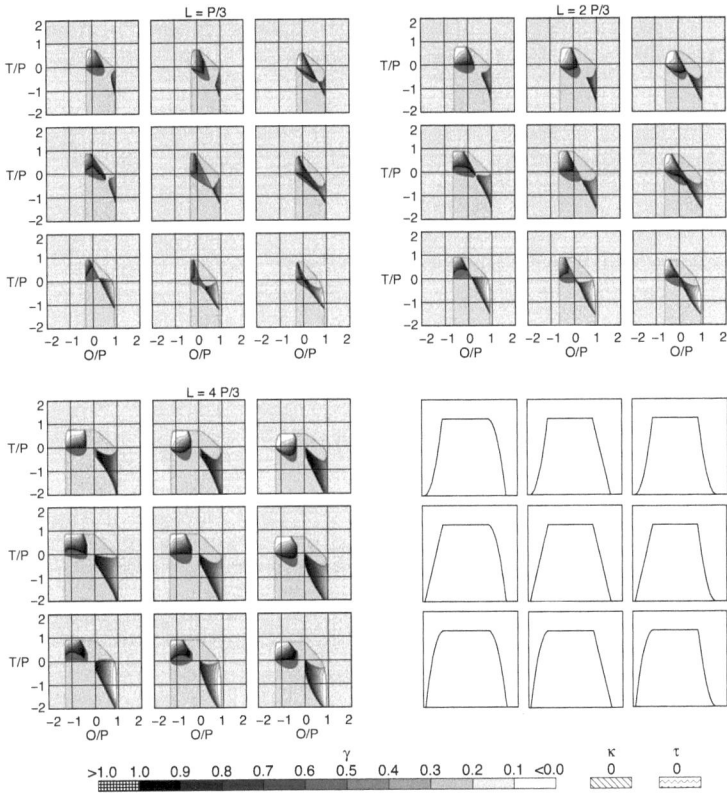

Figure 4.6: *Shown are γ_G values for different shapes of the signal u dependent on the ratio O/P and T/P for three different values of L/P. The upper left panel is for $L/P = 1/3$, the upper right for $L/P = 2/3$, and the lower left panel for $L/P = 4/3$, where P is the length of the rising as well as the falling phase. The different shapes are shown in the lower right and the corresponding equation (equation H.1) is given in the appendix H. The rows represent different η values (top to bottom: 0, 1 and 2) and the columns different ξ values (left to right: 0, 1 and 2). The individual figures are subdivided into a striped area where the weights will diverge ($\kappa_G \leq 0$), a zig-zag area where no overlap between both signals and the third factor exists and into a checked area that consists of γ_G values which, however, are beyond a meaningful range ($\gamma_G > 1$). The detailed gray shading represents γ_G values ($0 < \gamma_G \leq 1$), for which convergence is fulfilled.*

tor. Altogether it shows that for a general neuronally plausible signal shape u, the condition for asymptotic equivalence between temporal difference learning and differential Hebbian plasticity

CHAPTER 4 RELATION TO REINFORCEMENT LEARNING

with a *global* third factor is fulfilled for a wide parameter range covering all realistic relative timing intervals between state activations and global third factor.

Application: Linear network In this paragraph we show that we can reproduce the behavior of TD learning in a small linear network of neurons designed according to our algorithm. Obtained weights of the differential Hebbian plasticity neuron represent the corresponding TD value. It is known that in a linear TD learning system at the end of learning, values will follow an exponential function with a decay rate given by the discount factor γ. This is shown in panel (A). In panel (B) of this figure, we also investigate the assumption of a quasi-static process.

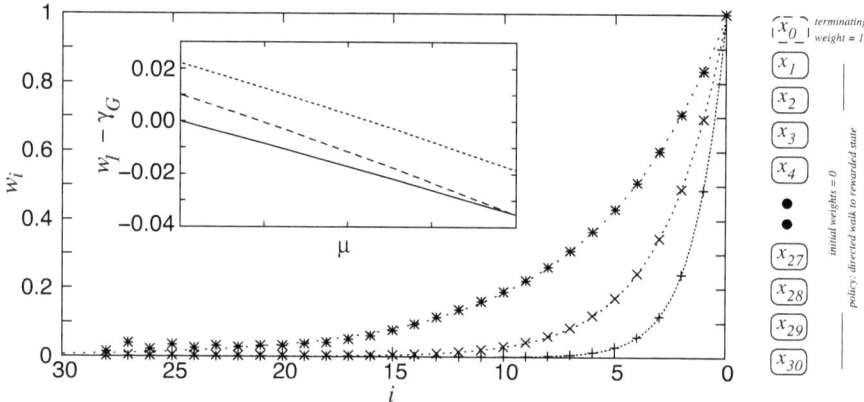

Figure 4.7: Shown are weights of a differential Hebbian plasticity neuron, where the arrangement of the states is shown on the right. On the left, the weights of the network and their corresponding exponential fit for three different γ_G values are plotted. The inset shows the dependence of the weights on the learning rate. The difference of the weight closest to the reward ($w_1 = \gamma_G w_0 = \gamma_G 1$) and the calculated γ_G value is plotted here and can be fitted by a logarithmic function $[f(x) \propto \log(1-x)]$. The γ_G values used are: (*, dotted) $\gamma_G = 0.835697$ $[S = 3000, T = 330, O = -220, L = 650]$, (x, dashed) $\gamma_G = 0.710166$ $[S = 3000, T = 300, O = -220, L = 650]$, (+, solid) $\gamma_G = 0.507729$ $[S = 3000, T = 300, O = -220, L = 550]$. The shape of the kernel used here is identical to the shape used in Figure 4.5, and the learning rate used for the main figure is 0.12.

Details of this simulation are as follows. The network consists of N states x which are connected to a neuron v which uses differential Hebbian plasticity. The modulatory signal is added by an additional neuron M. The states are indexed such that the state closest to the reward has index 1; hence, the reward has the index 0. The state structure is shown in Figure 4.7 right. At the beginning of learning, all weights are initialized to zero except the

4.2 GENERAL ANALYSIS

weight connected with the reward. Each trial begins with state N approaching the reward, at which a trial is terminated, thus, each state becomes active once.

The weights of the states connected to the differential Hebbian plasticity neuron are shown in Figure 4.7 A for three different γ_G values after learning. States indexed with higher numbers - hence, further away from the reward - have smaller weights and the relation $w_{i+1} = \gamma_G w_i$, where i indicates the distance to the reward holds for each γ_G value. This is indicated by an exponential fit. It also should be noted that the weights at states far away from the reward deviate from the exponential fit but only for the highest γ_G value. This is an effect caused by the finite number of states and at the same time by a γ^+ value which is higher than 1 (see last paragraph of section 4.2 for details).

In these system learning rates are usually in the range of 10^{-5} to 10^{-2} (Porr and Wörgötter, 2003b, 2007). The question arises whether in this range the assumption of a quasi-static process will hold. If it holds, we would expect that the weight closest to the reward (w_1) will reach exactly the value of γ_G after learning. In Figure 4.7 B the deviation from this expectation given by $w_1 - \gamma_G$ is plotted against the learning rate. As indicated by equation 4.16, the deviation increases with increasing learning rates, but remains small up to a rate of 10^{-1}, which is well in the range of useful learning rates. The actual shape of the curves is a consequence of different interacting processes depending, for example, on the total number of states (see technical discussion above) and others.

If looking at higher γ_G values, it is apparent that the effect of a finite number of states behaves antagonistically to the deviation caused by the increased learning rate, i.e. the weight after learning is shifted to higher values (independent of the learning rate). Therefore, if using higher γ_G values (or a smaller number of states), the simulated weight w_1 may be modified such that it will be identical to the calculated γ_G value, even if using finite learning rates larger zero. However, this will not correct the simulated γ_G value as such since the weights are then not arranged exponentially anymore (indicated by Figure 4.7).

Technical discussion When using a global third factor all constraints discussed in the beginning of this section hold. However, the application of an additional third factor allows handling stochastically uncertain environments in an easy way:

Stochastically uncertain environments. It is known that in stochastically uncertain environments, the TD values only converge with probability one when the learning rate decreases (Kushner and Clark, 1978; Dayan and Sejnowski, 1994). In our implementation, the signal M is constant. If it were instead implemented to diminish during repeated encounters with the same state, it would immediately incorporate the property of decreasing learning rates, too.

4.2.2 Local third factor

After having derived the equivalence between temporal difference learning and differential Hebbian plasticity with a *global* third factor, we will now repeat this derivation with a *local* third factor. The technical discussion at the end of this subsection will give insights into computational advantages compared to a global third factor.

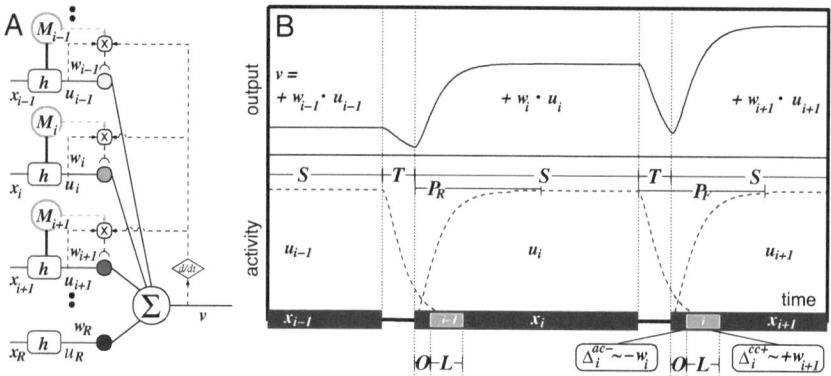

Figure 4.8: *The setup is shown in panel A and the signal structure in panel B. (A) Three states and the rewarded state converge on the neuron which learns according to equation 4.8. Each state x_i controls the occurrence of the modulatory factor M_i, which in turn will influence plasticity only at synapse w_i. This is different to Figure 4.4 where the modulatory factor influences all synapses. The states x will be active with increasing indices. (B) The lower part shows the states x_i, which have a duration of length S. We assume that the duration for the transition between two states is T. Above, the output v and the signals u are depicted. We additionally indicated the duration of the rising (P_E) and the falling phase (P_F) of the signals. Note that the duration for the plasticity and the output pathway are equal. Here u is given by $u(t) = \int_0^S (e^{-a(t-z)} - e^{-b(t-z)})\,dz$. The third factor M_i is released for the duration L after a time delay of O, which starts at the end of each state. This is different to the global third factor, which is initiated at the beginning of a state. The time the third factor is active is also shown in the lower part. For state x_i, the weight change contributions of the auto-correlation $\Delta^{ac\pm}$ and cross-correlation $\Delta^{cc\pm}$ are indicated.*

The local third factor M_i, in contrast to a global third factor, opens a time window, in which changes can occur only for its corresponding weight w_i (see Figure 4.8 A). This is indicated by the index i. Although this time window could be located anywhere depending on the input u_i, it should be placed at the end of the state x_i as it only makes sense if states correlate with their successor. Concerning the modulatory third factor M_i, we define its length as L, and the time period between beginning of M_i and the end of the corresponding state x_i as O. Similarly to the last subsection, this defines the boundaries of π. The four parameters (S, T, O, and L) are constant over states and are displayed in detail in Figure 4.8 B. Note that there is a significant difference in the definition of these parameters compared to the global third factor. There, the parameter O was defined between the beginning of the modulatory signal M and the *beginning* of the state x, whereas here it is defined towards the *end* of state x. This will always lead to

4.2 GENERAL ANALYSIS

a shift of $-T$ in our final equations in comparison to the equations we obtained using a global third factor. We will indicate the calculated κ, τ and γ values in this subsection with an index L (global).

Analysis of the differential equation Here the underlying equations 4.8, 4.9 and 4.10 can be found in table 4.1 on page 58. They differ from equation 4.5 only by the fact that the modulatory factor M has an index corresponding to its weight. Note that equation 4.9 is identical to equation 4.6.

The boundaries of the temporal path π are $t = S+O$ and $t = S+O+L$ for $ac-$ as we only have a time window at the end of the signal. Thus, $ac+$ as well as $cc-$ do not exist. Here we also use the same kernels for the plasticity and the output pathway, which allows us to simplify the integral in equation 4.17. Including the boundaries results in

$$\kappa_L(S,O,L) = \frac{1}{2}\left(u^2(S+O) - u^2(S+O+L)\right) \tag{4.33}$$

Thus the Δ_i^{ac}-function is equal to $\Delta_i^{ac-} = \tilde{\alpha}\,\kappa_L^{-}\,w_i$.

For the cross-correlation contribution we include the same boundaries discussed above into equation 4.22 as the modulatory factor M_i effects only the consecutive state. Thus, equation 4.21 is per definition zero. The remaining τ_L^{+}, here called τ_L, results in

$$\tau_L(S,T,O,L) = \int_{O-T}^{O+L-T} u(z+T+S)\,\dot{u}(z)\,dz \tag{4.34}$$

Note the time shift $-T$ in the integral boundaries compared to τ_G^{+} from equation 4.31. Here, the Δ_i^{cc}-function is identical to $\Delta_i^{cc+} = -\tilde{\alpha}\,\tau_L^{+}\,w_i$.

In accordance with the last subsection, both τ_L and κ_L depend on the actually used signal shape u and the values for the parameters S, T, O and L.

Analysis of the equivalence After we have calculated τ_L and κ_L and, more importantly, by finding that τ_L^{-} and κ_L^{+} do not exist when using a local third factor, this paragraph is simpler than the corresponding one in the first part of section 4.2. The equation after a single trial (equation 4.24) simplifies to

$$w_i \to w_i + \tilde{\alpha} - \kappa_L\,w_i + \tilde{\alpha}\,\tau_L\,w_{i+1}. \tag{4.35}$$

Directly after substituting $\alpha = \tilde{\alpha}\,\kappa_L$ and $\gamma_L = \tau_L/\kappa_L$, we arrive at the final equation

$$w_i \to w_i - \alpha\,w_i + \alpha\,\gamma_L\,w_{i+1} \tag{4.36}$$

without the use of equation 4.26. This allows us to drop two constraints, which we will discuss later in the technical discussion paragraph. Thus, likewise, if learning follows this local third factor differential Hebbian rule, weights will converge to the optimal estimated TD values. Furthermore, the analytics using a local third factor became more straightforward, and, as we will find out in the next paragraph, the convergence criterion is fulfilled for an even broader range of parameter (S, T, O, and L) values.

Analysis of the convergence Identically to the discussion of the convergence for the global third factor, we will investigate different signal shapes and the parameters (S, T, O and L), which also have an influence on the values of κ_L (equation 4.33) and τ_L (equation 4.34) and therefore $\gamma_L = \tau_L/\kappa_L$. Equation 4.35 gives us again two constraints: $\tau_L \geq 0$ and $\kappa_L \geq 0$ (where $\kappa_L = 0$ is allowed if $\tau_L = 0$). The situation with a local third factor is a great deal simpler than with a global third factor as we have to take only a single interval into account. If we again include only biophysically meaningful biphasic signal shapes it becomes evident that τ_L and κ_L can never be negative. The reason for τ_L can be found in equation 4.34 because our signal u is positive and so is the derivative of u at the onset of u, thus resulting in a non-negative value for τ_L. Concerning κ_L we also have to consider the phase of the third factor which is triggered in this case at the end of the signal x. As it starts to decay, $u(S+O)$ is always larger than $u(S+O+L)$ and therefore $\kappa_L > 0$. However, κ_L can become zero (indicated by striped regions in Figure 4.9 and Figure 4.10). This happens only if the third factor is shifted (due to the parameter O, see Figure 4.8 B for more details) to regions of the signal u, where the decay has not yet started ($O < -L$) or has already ended ($O > P$). The latter is the only case, where weights would diverge. If τ_L is zero (indicated by zig-zag regions in Figure 4.9 and Figure 4.10), weights will not diverge but simply stay zero. Note that this is a stronger constraint than $\kappa_L = 0$ as weights can not diverge, in the worst case they do not change at all. This is the reason why regions with $O < -L$ will not diverge although κ_L is equal to zero. Discount factors, which are represented by γ_L values exceeding 1, are likewise not considered in our further discussion.

In the previous subsection, we analyzed the γ_G values for the function used in Figure 4.8, so we will do the same for γ_L values here, too. Identical to Figure 4.5, Figure 4.9 shows the γ_L value in 3 different panels. In each panel we varied the parameters O and T from minus $2P$ to plus $2P$ where again $P = P_E = P_F$ is the time the signal u needs to reach the maximum. In each of the panels, we plot γ_L values for a particular value of L.

We additionally investigate the parameter regions, where our weights converge to the values predicted by temporal difference learning for our Taylor expanded signal shapes given by equation H.1. We find that, as discussed before, there exists only one region per panel in Figure 4.10 which gives us an infinite γ_L value, resulting in divergent weights.

Here we also use the extracted information about the regions, where γ_L is either convergent or divergent (see appendix H) and compare the resulting Figure H.8 with the results of Figure 4.10. We find that both figures match to each other as we have used the same derivations in the appendix as for Figure 4.10. And even if you use more general kernel functions, like the exponential function used for Figure 4.9, both figures still match quite well, especially in the regions, where the system diverges and where it stays constant. This is consistent with the findings for the global third factor.

4.2 GENERAL ANALYSIS

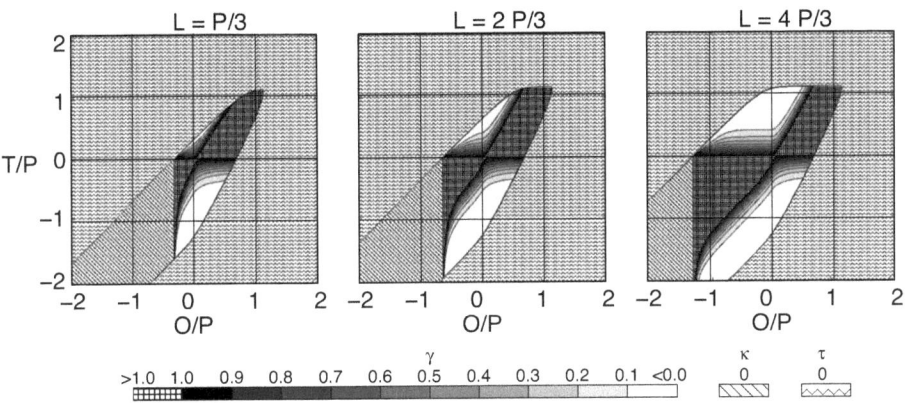

Figure 4.9: Shown are γ_L values dependent on the ratio O/P and T/P for three different values of L/P (1/3, 2/3, and 4/3). Here P is the length of the rising as well as the falling phase. The shape of the signal u is identical to the one used in figure 4.8 and is given by $u(t) = \int_0^S (e^{-a(t-z)} - e^{-b(t-z)})\,dz$ with parameters $a = 0.006$ and $b = 0.066$. The individual figures are subdivided into a striped area, where the weights will diverge ($\kappa_L = 0$, a zig-zag area where no overlap between both signals and the local third factor exists and a checked area that consists of γ_L values which, however, are beyond a meaningful range ($\gamma_L > 1$). The detailed gray shading represents γ_L values ($0 < \gamma_L \leq 1$), for which convergence is fulfilled.

The different frames show clearly that the area of convergence changes only gradually and the area as such is increasing with increasing duration of the local third factor. Altogether it shows that for a general neuronally plausible signal shape u the condition for asymptotic equivalence between temporal difference learning and differential Hebbian plasticity with a *local* third factor is fulfilled for an even wider parameter range compared to a global third factor.

Application: Linear network In this paragraph we show that we can reproduce the behavior of TD learning in a small linear network with two terminal states (see Figure 4.11 left). This is done with a network of neurons designed according to our algorithm with a local third factor. Obtained weights of the differential Hebbian plasticity neuron represent the corresponding TD values. It is known that in a linear TD learning system with two terminal states (one is rewarded, the other not) and a γ value close to 1, values at the end of learning will represent the probability of reaching the reward state starting at the corresponding state (compare Sutton and Barto (1998)). This is shown in Figure 4.11 including the weight development.

Technical discussion When using a local third factor we can, as before, also handle stochastic uncertain environments (letting the modulatory factors decay to zero). However, we can

74 CHAPTER 4 RELATION TO REINFORCEMENT LEARNING

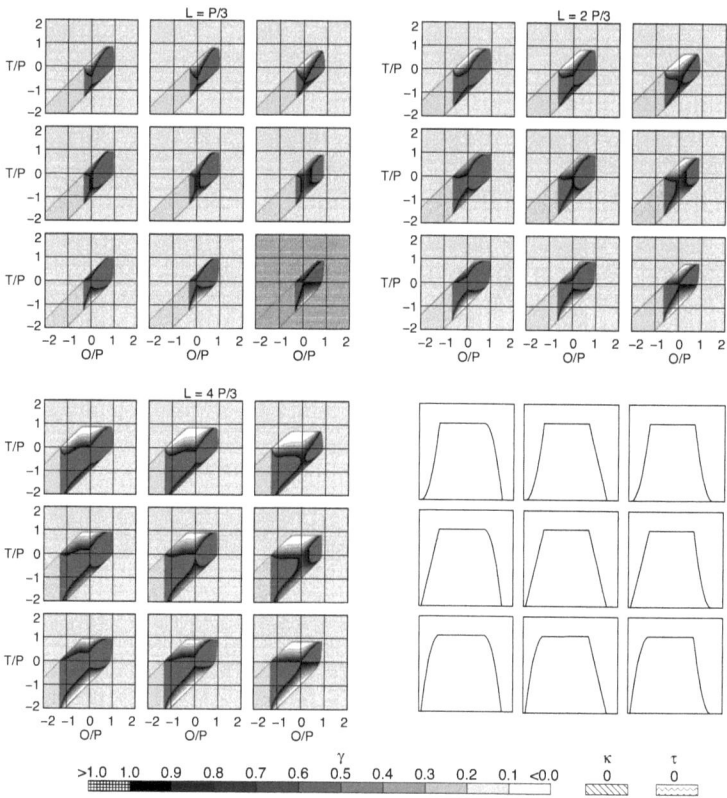

Figure 4.10: *Shown are γ_L values for different shapes of the signal u dependent on the ratio O/P and T/P for three different values of L/P. The upper left panel is for $L/P = 1/3$, the upper right for $L/P = 2/3$, and the lower left panel for $L/P = 4/3$, where P is the length of the rising as well as the falling phase. The different shapes are shown in the lower right and the corresponding equation (equation H.1) is given in the appendix H. The rows represent different η values (top to bottom: 0, 1 and 2), and the columns different ξ values (left to right: 0, 1 and 2). The individual figures are subdivided into a striped area, where the weights will diverge ($\kappa_L = 0$), a zig-zag area, where no overlap between both signals and the third factor exists ($\tau_L = 0$) and into a checked area that consists of γ_L values which, however, are beyond a meaningful range ($\gamma_L > 1$). The detailed gray shading represents γ_L values ($0 < \gamma_L \leq 1$), for which convergence is fulfilled.*

now drop two constraints as equation 4.35 does not dependent on the preceding state anymore. First, we can lift the constraint which came with the number of states and the actual starting position. As the weight change does not depend on the previous state, the starting position

4.2 GENERAL ANALYSIS

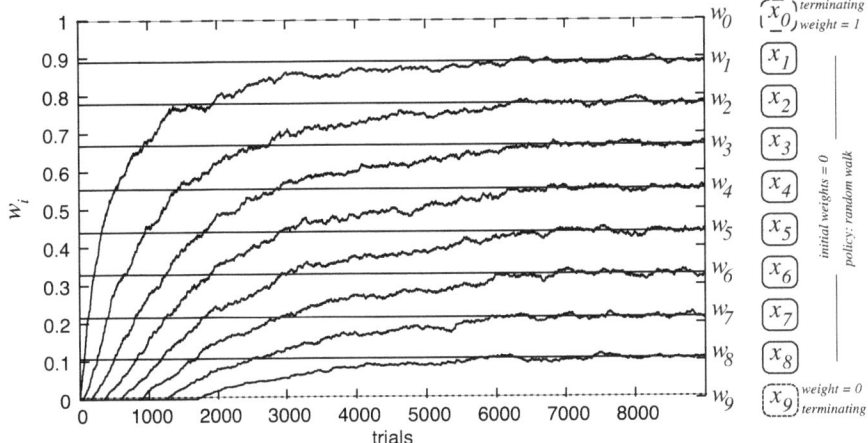

Figure 4.11: This figure shows the weight development and the linear state arrangement with two terminating states. The weights which after convergence correspond to the states depicted on the right are distributed uniformly (compare Sutton and Barto (1998)). The lines represent the mean of the last 2000 weight values of each state and coincide with the TD values we would get. The signal shape is given by $u(t) = \int_0^S (e^{-a(t-z)} - e^{-b(t-z)}) \, dz$ with parameters $a = 0.006$, $b = 0.066$ and $S = 10000$ which result in $P = 1200$. Furthermore is $O = 1/20\,P$, $L = P$, $T = 0$ (which yields $\gamma_L \simeq 1$) and the learning rate 0.01.

is treated identical to the subsequent states. Second, now the system is strictly Markovian, which allows us to use this realization directly without redefining the state space. Thus, the linear network with two terminal states and, most importantly, with a random policy would have developed different weights when using the global third factor without additional modifications.

4.2.3 Different time scales: VOT plasticity

In section 2.1 we showed that the auto-correlation is adjustable by the usage of different time scales, i.e. by using different kernels or rather kernel parameters that change the output trace. The question which now arises is whether we can use the same idea and implement reinforcement learning by using different time scales.

The setup is depicted in Figure 4.12 A where kernels for the plasticity- and output pathway are different. The corresponding signal structure is shown in Figure 4.4 B having different signal shapes for plasticity- and output signals. The boundaries of π are defined by the rising $P_{v,E}$ and falling $P_{v,F}$ period of the output signals u_v where v stands for the different kernel parameters and their consequences. Using the results from section 2.1 we will use a smaller time scale for the output kernels, i.e. shorter rising and falling time, which will lead to a negative

CHAPTER 4 RELATION TO REINFORCEMENT LEARNING

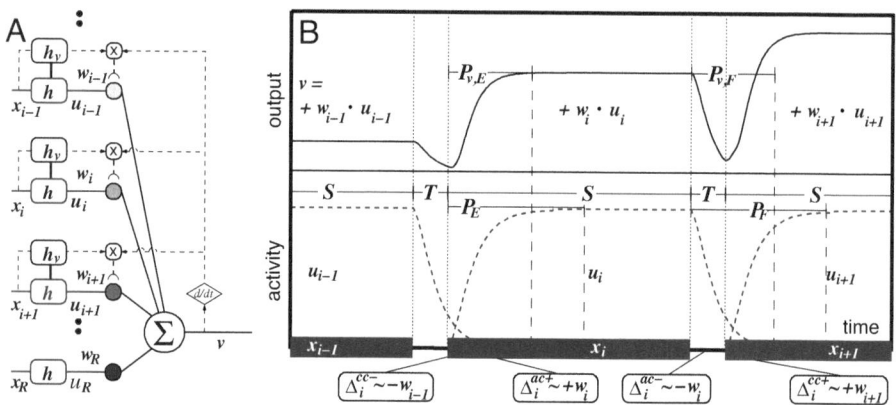

Figure 4.12: *The setup is shown in panel A and the signal structure in panel B. (A) Three states and the rewarded state converge on the neuron which learns according to equation 4.11. By contrast to Figure 4.4 and 4.8, no third factor is used. The states x will be active with increasing indices. (B) The lower part shows the states x_i which have a duration of length S. We assume that the duration for the transition between two states is T. Above the output v and the signals u are depicted. We additionally indicated the duration of the rising (P_E and $P_{v,E}$) and the falling phase (P_F and $P_{v,F}$) of the signals and the output respectively. Note that different to Figure 4.4 and 4.8 the kernel parameters for plasticity and output are different. This leads to difference in the duration of rising and falling phase indicated by two dashed lines. Here u is given by $u(t) = \int_0^S (e^{-a(t-z)} - e^{-b(t-z)})\,dz$. For state x_i the weight change contributions of the auto-correlation $\Delta^{ac\pm}$ and cross-correlation $\Delta^{cc\pm}$ are indicated.*

auto-correlation contribution. Using this we will calculate κ, $\tau^\pm$ and γ which will be indicated by T (time scale).

Analysis of the differential equation The underlying equations 4.8, 4.9 and 4.10 can be found in table 4.1 on page 58. They are identical to equations 2.5 and 2.13 of the last paragraph of section 2.1 covering differential Hebbian plasticity with different time scales.

The boundaries of the temporal path π are $t = 0$ and $t = P_{v,E}$ for $ac+$ and $t = S$ and $t = S + P_{v,F}$ for $ac-$. Here, the integral of equation 4.17 is not in general solvable, thus we only include the boundaries which results in

$$\kappa_T(S, v) = \int_0^{P_{v,E}} u_i(z)\,\dot{u}_{v,i}(z)\,dz$$
$$+ \int_S^{S+P_{v,F}} u_i(z)\,\dot{u}_{v,i}(z)\,dz$$
$$= -\left(\kappa_T^+(v) + \kappa_T^-(S,v)\right). \tag{4.37}$$

4.2 GENERAL ANALYSIS

Thus, similar to the global third factor, the Δ_i^{ac}-function splits into $\Delta_i^{ac+} = \tilde{\alpha}\,\kappa_T^+\,w_i$ and $\Delta_i^{ac-} = \tilde{\alpha}\,\kappa_T^-\,w_i$.

For the cross-correlation contribution, we include the same boundaries discussed above into equations 4.22 and 4.21, which leads to

$$\tau_T^-(S,T,v) = -\int_0^{\max(P_{v,F}-T,0)} u(z)\,\dot{u}_v(z+S+T)\,dz \tag{4.38}$$

$$\tau_T^+(S,T,v) = \int_0^{P_{v,E}} u(z+T+S)\,\dot{u}_v(z)\,dz \tag{4.39}$$

where τ_T^- is equal to zero if the transition time T is greater than $P_{v,F}$. Thus, in general the Δ_i^{cc}-function splits into $\Delta_i^{cc-} = -\tilde{\alpha}\,\tau_T^-\,w_i$ and $\Delta_i^{cc+} = \tilde{\alpha}\,\tau_T^+\,w_i$. Both $\tau_T^\pm$ and $\kappa_T^\pm$ depend on the actually used signal shapes u and u_v and the values for the parameters S and T.

Analysis of the equivalence As we have calculated $\tau_T^\pm$ and $\kappa_T^\pm$ and found out that for some parameter values ($T > P_{v,F}$) τ_T^- does not exist, this paragraph is a mixture of the corresponding paragraphs of section 4.2 and subsection 4.2.2. If $\tau_L^- = 0$, we simplify the calculations according to subsection 4.2.2. On the other hand, if $\tau_L^- \neq 0$, we need to stick to the more complex derivation used in subsection 4.2.1. Similar to the preceding sections, if learning follows this differential Hebbian rule with different time scales of plasticity and output, weights will converge to the optimal TD values. The convergence properties (see next paragraph) are even better than for the local third factor, however, the Markov property is not always fulfilled.

Analysis of the convergence Here we will cover the conditions for the signal shapes u and u_v and the parameters (S and T), which will lead to the demand that γ_T should be between zero and 1 ($0 < \gamma_T \leq 1$) and that κ_T should be strictly positive ($\kappa_T > 0$).

In the previous sections, we analyzed the $\gamma_{G/L}$ values for the function used in Figure 4.3, so we will do the same for γ_T values here. For this we have to define the output kernels different from the plasticity kernels according to equations 2.5 and 2.13. We do this by using an ρ value which scales the time of output kernel relative to the plasticity kernel (see equation 2.16). An ρ value of infinity relates to a δ-function, thus to the S&B model. Figure 4.13 left shows that even the strict demand that γ_T needs to be bounded between 0 and 1 holds for all possible T values given an ρ value greater 1. This corresponds to the fact that the output kernel function h_v is narrower than the plasticity kernel h. On the right side of Figure 4.13, we plot the γ_T value for $\rho \to \infty$, hence for the S&B model (see appendix H.7 for an analytical solution of γ_T for the S&B model).

To extend these considerations to more general shapes we use equation H.1 to calculate γ_T analytically (see appendix H for the definition of the signal shape and appendix H.6 for the analytical calculation). The results are then plotted in Figure 4.14 with respect to ρ and T for nine different input functions which are shown in the top part of each panel. Additionally we show in appendix H.6 that κ_T is always positive. This explains why Figure 4.14 and 4.13 do not have regions for which the system is divergent.

CHAPTER 4 RELATION TO REINFORCEMENT LEARNING

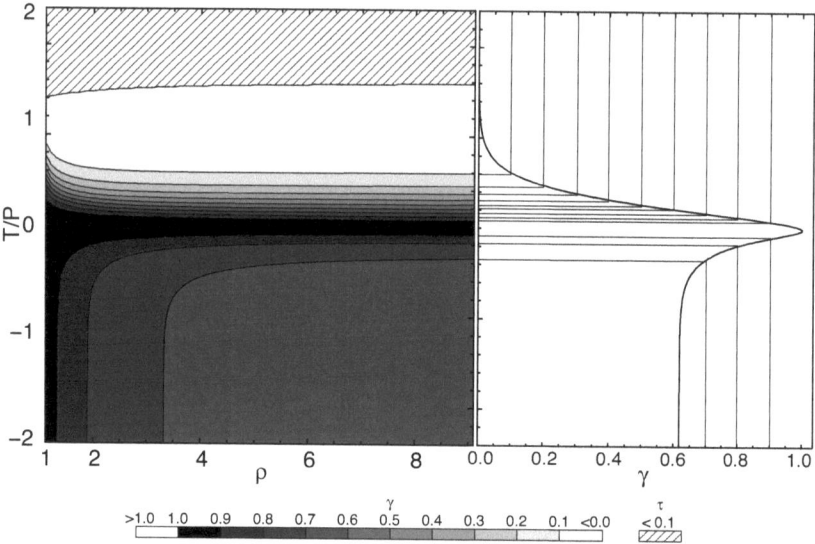

Figure 4.13: *Shown are γ_T values dependent on the ratio T/P and ρ (see equation 2.16). Here $P = P_{v,E} = P_{v,F}$ is the length of the rising as well as the falling phase of the output. The shape of the signal u is identical to the one used in Figure 4.3 and Figure 4.12 and is given by $u(t) = \int_0^S (e^{-a(t-z)} - e^{-b(t-z)})\,dz$, with parameters $a = 0.006$ and $b = 0.066$. The detailed gray shading represent γ_T values $(0 < \gamma_T \leq 1)$ for which convergence is fulfilled within a meaningful range. In the striped regions we have $\tau_T < 0.1$, thus almost no overlap. On the right we plotted the γ_T for $\rho \to \infty$ which resembles the S&B model with a delta pulse for the output pathway.*

Both figures (Figure 4.13 and Figure 4.14) indicate that when using two different time scales, the convergence is guaranteed for all T parameters as long as the time scale of the output is smaller than that of the plasticity.

Application: Linear network Depending on the transition time T between two states this network can produce results which resemble either the linear network with a random policy (compare to subsection 4.2.2) or only the linear network with a gradient policy (compare to subsection 4.2.1).

Technical discussion Here we have a limited set of parameters (ρ and T) that can be varied, and the last paragraph revealed that for $\rho > 1$ all T values lead to non-divergent weight values. Additionally, which is also a favorable result, it restricts the corresponding γ_T value to 1. If we also set T to be large enough, i.e. $T > P_{v,F}$, the constraints related to the Markov property would be released, identically to the local third factor. For overlapping states, however, the

4.2 GENERAL ANALYSIS

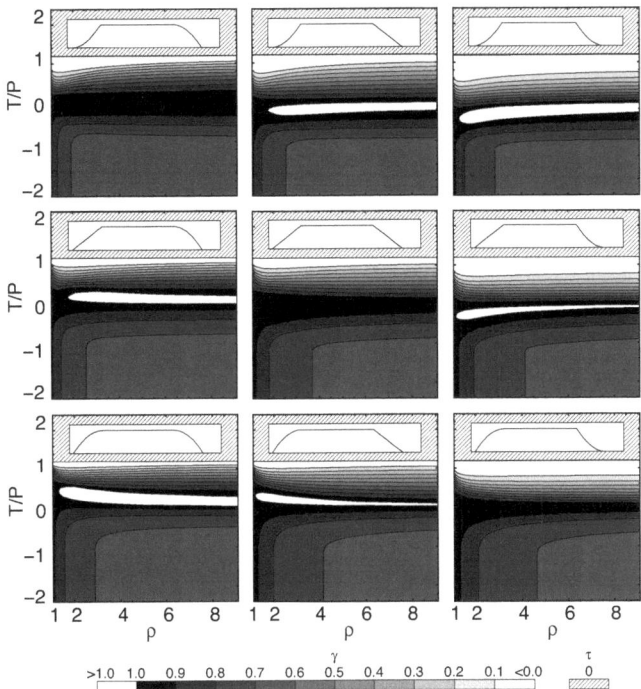

Figure 4.14: *Shown are γ_T values for different shapes of the signal u dependent on the ratio T/P and ρ (see equation 2.16). The different shapes are shown in the top part of each panel and the corresponding equation (equation H.1) is given in appendix H. The rows represent different η values (top to bottom: 0, 1 and 2) and the columns different ξ values (left to right: 0, 1 and 2). The detailed gray shading represent γ_T values ($0 < \gamma_T \leq 1$), for which convergence is fulfilled. In the striped regions we have $\tau_T = 0$, thus no overlap. Regions that result in γ values greater 1 are only existent for non-symmetrical signal shapes (e.g. $\eta = 0$ and $\xi = 2$).*

system stays non-Markovian. Thus, this mechanisms is a mixture of the global and the local factor as it combines the properties of both.

How is this method of VOT plasticity related to a third factor. In principle, it acts similarly to such a factor as a smaller time scale for the output pathway restricts plasticity to a smaller time window, i.e. smaller than the intrinsic time window given by the rising and falling phase of the plasticity kernels. Hence, a third factor with the functional characteristics of the output kernels would achieve identical results and properties.

4.3 Discussion

The TD rule has become the most influential algorithm in reinforcement learning, because of its tremendous simplicity and proven convergence to the optimal value function (Sutton and Barto, 1998). It was successfully transferred to control problems, too, in the form of Q- or SARSA learning (Watkins and Dayan, 1992; Singh et al., 2000), which use the same algorithmic structure, while maintaining similar advantageous mathematical properties (Watkins and Dayan, 1992).

Relation to other work

In this study we have shown that TD(0) learning and differential Hebbian plasticity either modulated by a third factor or by using different time scales for plasticity and output pathway are asymptotically equivalent under certain conditions. This proof relies only on commonly applicable, fairly general assumptions, thus rendering a generic result not constraining the design of larger networks. It has long been suspected that RL would, in neuronal tissue, have to rely on the use of a third factor in a Hebb rule (Schultz, 1998) and several earlier results have pointed to the possibility of an equivalence between reinforcement and correlation-based learning. Izhikevich (2007) solved the distal reward problem using a spiking neural network, yet with fixed exponential functions (Gerstner et al., 1996) to emulate differential Hebbian characteristics. His approach is related to neurophysiological findings on spike-timing-dependent plasticity (STDP, Markram et al. (1997)). Each synapse learned the correlation between conditioned stimuli and unconditioned stimuli (e.g. a reward) through STDP and a third signal. Furthermore, Roberts et al. (2009) showed that asymmetrical STDP and temporal difference learning are related. In our differential Hebbian learning model, in contrast to the work described above, STDP emerges automatically because of the use of the derivative in the postsynaptic potential (equation 2.25). The relation between STDP and differential Hebbian learning and its asymptotic equivalence when using serial states was discussed in Roberts (1999). Rao and Sejnowski (2001) showed that using the temporal difference will directly lead to STDP, but they could not provide a rigorous proof for the equivalence. Recently it has been shown that the online policy-gradient RL-algorithm (OLPOMDP) developed by Baxter et al. (2001) can be emulated by spike-timing-dependent plasticity (Florian, 2007), however, in a complex way using a global reward signal. On the other hand, the observations reported here provide a rather simple, equivalent correlation-based implementation of TD and support the importance of three-factor learning for providing a link between conventional Hebbian approaches and reinforcement learning.

Relation to function value approximation

One drawback of reinforcement learning algorithms, like temporal difference learning, is their use of discrete time and discrete non-overlapping states. In real neural systems, time is continuous and the state space can only be represented by the activity of neurons, many of which will be active at the same time and for the same "space". This creates a rather

4.3 DISCUSSION

continuous state space representation in real systems. In order to allow for overlapping states or for generalizing over a wider range of input regions, RL algorihtms are usually extended by value function approximation methods (Sutton and Barto, 1998). However, while biologically more realistic (Tamosiunaite et al., 2008), this makes initially elegant RL algorithms often quite opaque, and convergence can many times not be guaranteed anymore (Tsitsiklis and Van Roy, 1997; Wiering, 2004). Here we are not concerned with function approximation, but instead address the question of how to transform an RL algorithm (TD learning) to continuous time using differential Hebbian plasticity and remaining fully compatible with neuronally plausible operations. However, with the algorithm presented here clearer and more rigorous convergence proofs could be developed. Only a few other approaches to formulate RL in continuous time and space exist (Baird, 1993; Doya, 1996, 2000), however lacking biological motivation. In particular Baird (1993) extended Q learning by the "advantage updating" method and Doya (2000) performed the transformation from a discrete sum to a continuous integral for the calculation of the return R. In his case every value function V consists of a state representation and a corresponding weight. These weights need to be adjusted in order to let the δ error converge to zero. This is done by a gradient descent algorithm which results in an update rule that demands a weight derivative, which is difficult to emulate in a biologically realistic way.

Historical remark

It is interesting from a historical viewpoint that Sutton and Barto switched from a setup similar to that presented in this chapter to a serial compound representation (see section 2.3) when switching to temporal difference learning (Sutton and Barto, 1990). The main concern they had with this setup is the simultaneous occurrence of a stimulus and a rewarded stimulus. In such a case the weight of the stimulus converges to a value that counterbalances the weight of the reward. This happens for every plasticity rule which has a negative auto-correlation contribution. Thus, the setup presented here is also affected by this problem. There are two simple arguments why we should not be concerned. First, if we do not allow weights to become negative, the reward value, which is mostly positive, can not be counterbalanced. Second, although many different stimuli occur at the same time the reward is presented, these stimuli do not exactly fit with respect to timing and temporal development. Therefore, although the auto-correlation is negative, the weight change could be positive.

Remark concerning closed-loop systems

It is also a question how the parameter O, which represents the onset of the third factor, is implemented in behaving systems, in particular if O is negative. This requires the system to know when the next state is reached, hence it would need another algorithm that anticipates the timing of upcoming states. On the other hand Figures 4.5, 4.6, 4.9, and 4.10 show that the areas of convergence are still sufficiently large in the positive range of O values.

Relation of the third factor to neuromodulators

In this part of the thesis we are mainly concerned with showing the formal equivalence between TD and differential Hebbian plasticity. Possible links to biophysical mechanisms play a minor role here. Nonetheless, one could consider neuromodulators for the role of the third factor M. The required reliability of timing, however, makes it unlikely that Dopamine could take on this role as the timing of these signals does not seem to be reliable enough (Redgrave and Gurney, 2006), although Pawlak and Kerr (2008) could show that LTP in the Striatum only emerges in the presence of Dopamine. The attributed, albeit still much discussed, role of the dopaminergic responses from neurons in the Substatia Nigra (pars compacta) or the Ventral Tegmental Area (VTA) as possibly representing the δ error of TD learning (Schultz et al., 1992; Montague et al., 1996) is, thus, neither questioned nor supported by the results presented here. A very good alternative for the role of a well-timed third factor, however, seems to be the response characteristic of the cholinergic tonically active neurons (TAN) releasing the neuromodulator Acetylcholine. Their response, which is a reduction of activity, is exceedingly well timed and occurs together with conditioned stimuli (Graybiel, 1998; Morris et al., 2004). The fact that TAN's cease to fire, would require an additional inversion to make this compatible with our M factor, but when considering possible disinhibitory effects, this should not pose a fundamental problem.

It is also important that we were able to show that our algorithm is indeed stable across a wider range of possible biological signals as different temporal profiles exist e.g. for synapse and channel activation (compare AMPA vs. NMDA characteristics). This is required as it is not clear at this point in time - as discussed above - which signals are involved in any three-factor learning and this might also depend on the considered cell type and brain structure.

We also found that independent of which kind of third factor one uses, there exists a wide parameter range in which differential Hebbian plasticity becomes asymptotically equivalent to temporal difference learning. We could also show that not only the analytical treatment of the local third factor is simpler but also the convergence is stronger as compared to a global third factor. Furthermore, constraints concerning the number of states and the non-Markovian property are lifted as the local third factor only correlates states with following states and not with preceding ones.

On the other side, it is difficult to find biological counterparts for a local third factor. The main difference between a global and local third factor is the origin of the afferents to the neurons producing this signal. In the case of the global third factor, the output of the considered neuron, which could be represented by neurons in the Striatum, is the driving force of the release of neuromodulators. By contrast, for the local third factor it is the input. The origin of this input is either cortical or hippocampal. Within the Striatum TANs are favorable substrates producing the third factor. However, these TANs are interneurons and their input is mainly provided by other neurons of the Striatum. This would favor the global aspect of the third factor.

4.3 DISCUSSION

Importance of the negative auto-correlation contribution

The last section of this chapter showed clearly that any given plasticity rule that uses the proposed general setup and has a negative auto-correlation contribution is able to emulate temporal difference learning.

Chapter 5
Discussion and Outlook

In this thesis we focused mainly on the analysis of the auto- and cross-correlation of a synaptic connection while it is changing under a Hebb-like plasticity rule. The cross-correlation describes the correlation of the corresponding input with inputs of other weights. By contrast, the auto-correlation depicts the correlation of the corresponding input with itself. Hence, it becomes visible when only the corresponding input drives the output. This provides a general theoretical framework which allows us to make predictions about the overall weight development in all Hebb-like models. This theoretical framework led to insights with which a setup was developed that made it possible to prove the long suspected equivalence between differential Hebbian plasticity and temporal difference learning. Additionally we developed an analytical solution which describes the weight development in time of arbitrary many plastic synapses with non-stationary input patterns.

Relation to classical conditioning
The models investigated in this thesis are centrally related to classical conditioning. For instance, in section 2.1 we presented in detail the S&B model (Sutton and Barto, 1981) which was originally used to describe classical conditioning. It was the first real-time computational model that could explain data from animal experiments. The S&B model makes all the same predictions as the Rescorla-Wagner model (Rescorla and Wagner, 1972) which was, however, a trial-level model. Real-time models describe the temporal development step-by-step, by contrast trial-level models only take trial (one trial consists of many time steps) relevant information into account. For instance, they are only interested whether there was a second stimulus and not, in case the second stimulus appeared, at which step in time it occurred. Because of this, the S&B model, as well as the here proposed VOT model (see section 2.1), can explain, among others, some of the inter-stimulus interval effects, however, not all of them. This was one of the reasons that Sutton developed the temporal difference model (see section 2.3), which was instrumental in shaping the field of Reinforcement Learning (Sutton and Barto, 1998). The TD rule had problems predicting the S-shaped weight development found in the data of animal experiments. There the slope of the weight development increases with increasing weight before converging to the final value. In the same year, however, Klopf

Name	Output: v	Rule: $\dot{w}_1$	Δw^{ac}	Δw^{cc}	Comment
S&B	$w_0\,x_0 + w_1\,x_1$	$u_1\,\dot{v}$	<0	$-$	$cc=0$ for negative T
ISO	$w_0\,u_0 + w_1\,u_1{}^*$	$u_1\,\dot{v}$	$=0$	$-$	cc anti-sym., numerical unstable
VOT	$w_0\,u_{v,0} + w_1\,u_{v,1}$	$u_1\,\dot{v}$	$-$	$-$	different time scales
Hebb	$w_0\,u_0 + w_1\,u_1$	$u_1\,v$	>0	>0	always divergent
TD	$w_1\,x_1$	$u_1\,r + u_1\,\dot{v}$	<0	>0	$cc=0$ for negative T
ICO	$w_0\,u_0 + w_1\,u_1{}^*$	$u_1\,\dot{u}_0$	$=0$	$-$	cc anti-sym., numerical stable
ISO3	$w_0\,u_0 + w_1\,u_1{}^*$	$u_1\,\dot{v}\,\overline{R}$	$-$	$-$	3-factor diff. Hebb

Table 5.1: *Overview over all single synapse plasticity rules discussed in this thesis. The asterisk * depicts identical equations within one column. The auto-correlation contribution is abbreviated with Δw^{ac} and the cross-correlation contribution with Δw^{cc}. A hyphen $-$ indicates that the value is not restricted to neither the positive nor the negative range.*

presented a model (Klopf, 1988), which was an extension of the Drive-Reinforcement model (Klopf, 1982), that made all the same predictions as the TD model. In his model, Klopf introduced a derivative in the input and added the weight into the eligibility trace. The latter results in an S-shaped weight development. For additional models that predict certain aspects of classical conditioning see the review from Balkenius and Morén (1998).

Overall weight change: Auto- and cross-correlation

In classical conditioning there are essentially two stimuli, namely the conditioned stimulus which comes earlier and the unconditioned stimulus which follows. Therefore we analyzed in chapter 2 and 3 the weight development of a single plastic synapse in the presence of two consecutive spikes that were convoluted by kernel functions. Although the detailed temporal development had interesting features, we were mainly interested in the final weight after such an event. In particular, we focused on the question which contributions (auto- and/or cross-correlation) the weight change consisted of. For a better overview we summarized the rule, the output and the auto/cross-correlation contributions for all discussed plasticity rules in table 5.1. We presented the cross-correlation in weight change curves which were always plotted with respect to the temporal difference of the two consecutive spikes. Here we found different types of curves; symmetrical for Hebbian plasticity, anti-symmetrical for differential Hebbian plasticity and combinations of these extreme cases (e.g. for differential Hebbian plasticity with a third factor). The second contribution, namely the auto-correlation, can be found in these curves at $T=0$. We distinguished between three different classes: 1) negative, 2) zero or 3) positive auto-correlation.

<u>1) negative:</u> The first class which has negative auto-correlation was used in chapter 4. There it was important that the weight converges if its corresponding input is paired subsequently with another later input, for instance with the reward. The reason for this convergence is that the overall weight change results in a difference equation (see equation 2.17 and chapter G) which only converges for negative auto-correlations.

Tropism-like
ISO
ICO
ISO3
VOT
Delayed Rewards-like
S&B+
TD+
Actor-Critic
Q/SARSA learning
neuronal Actor-Critic
neuronal mix of Q- and SARSA learning

Table 5.2: *Categorization of models discussed in this chapter in the context of closed-loop systems into tropism-like and delayed reward-like rules. TD+ and S&B+ indicates that, in order to be used in closed-loop systems, the TD and the S&B model, respectively, have to be extended.*

2) zero: To achieve final stability of weights when the second (later) stimulus is avoided, the algorithmic class with zero auto-correlation is of interest. As both the cross-correlation and the auto-correlation contribution are zero, the overall weight change is "per definition" zero.

3) positive: The class with positive auto-correlation is the worst as it drives an already positive weight to infinity even without an additional input. If additionally the cross-correlation is strictly positive, as is the case for Hebbian plasticity, the weight changes only to more positive values and wrongly learned correlations can never be unlearned again.

Closed loop with zero auto-correlation

Positive auto-correlations, even if they have numerical origins like in the ISO rule, are problematic as they cause the weights to diverge. Therefore, we put essential effort in chapter 2 and also in section 3.1 to find plasticity rules that fall into the class with a zero auto-correlation. This becomes particularly important in the context of behaving systems, i.e. in systems, where the output of our small neuronal network is influencing the inputs the network receives. Here, diverging weights would lead to nonsensical output values, thus to unwanted behavior. In earlier work of Porr and Wörgötter (e.g. Porr and Wörgötter (2003a,b, 2006, 2007)) closed-loop systems were equipped with plasticity rules also found in this thesis. For instance in Porr and Wörgötter (2003a), a Braitenberg vehicle (Braitenberg, 1984) learned to avoid obstacles using the ISO rule (homosynaptic differential Hebbian plasticity) or in Porr and Wörgötter (2006) a mechanical arm learned using the ICO rule (heterosynaptic differential plasticity) to anticipate disturbing impulses in order to keep its position. For the latter, we developed in Kolodziejski et al. (2006, 2007) improvements to overcome problems with different impulse strengths. In the closed-loop context the difference between the numerically instable ISO rule and the stable ICO rule becomes critical for the ISO rule which leads to weight configurations that can not solve the task anymore. To this end, an extensive comparison was conducted in Porr and Wörgötter (2006) where it was found that the ICO rule solves the tasks by far better than the ISO rule. The reason is the intrinsic absence of the auto-correlation contribution in the ICO rule. This leads to a perfect convergence when no second stimulus is present, i.e. when the system avoids the second stimulus. Therefore this rule should be used whenever a stimulus needs to be avoided or compensated, e.g. in an obstacle-avoiding task or, in the field of engineering, for thermostat-like tasks, i.e. whenever the systems needs to converge to a set point. Earlier we mentioned that a negative auto-correlation is useful when an agent wants to approach a rewarding stimulus and by contrast an auto-correlation with zero value should be used when an agent wants to avoid a punishing stimulus. However, Porr and Wörgötter found a trick to use the stimulus-avoiding condition also in a positive tropism task, i.e. a task in which the agent needs to reach (approach) the stimulus. For this, the authors used the symmetry of two closely attached sensors when approaching a target (e.g. circular food blob). The difference which defines the actual stimulus between those two sensors is zero as soon as the agent approaches the circle perpendicularly, hence avoiding this difference stimulus. With this differential stimulus setup, food retrieval tasks were studied in Porr and Wörgötter (2006, 2007). In order to classify different rules into tropism-like or delayed reward-like (see below) closed-loop rules, we list them in table 5.2.

Multi-synapse systems and their relation to closed-loop systems

In natural environments an agent usually has to deal with more than one stimulus at a given time. Therefore in chapter 3, in particular in section 3.2, we extended the analytical considerations to many plastic synapses that change at the same time by means of a general linear Hebbian plasticity rule. We found, for the first time, a complete analytical solution for continuously varying inputs. We also provided appropriate approximations as the complete solution is difficult to compute. Usually, if input patterns are stationary, we can use well-known methods, like taking the eigenvalues of the covariance matrix (Dayan and Abbott, 2001), in order to predict the weight development. As the environment for behaving, hence self-stimulating, systems is usually complex, the input patterns in general can not be simplified and we need to deal with realistic non-stationary input patterns. Equipped with the tools derived in section 3.2 we are now able to face biologically realistic neuronal setups, thus predicting their weight development. An interesting starting point would be to predict input patterns that lead to either stable, periodic or chaotic weight dynamics. Also the final weight distribution is of interest as it would allow us to compare those results to already existing statistical analysis of experimentally measured distributions of synaptic connections (Barbour et al., 2007). Another direction would be to further investigate along the lines of the relation among spike-timing-depending plasticity (STDP) and Hebb-like plasticity rules (see chapter 2, Roberts (1999); Porr et al. (2004); Saudargiene et al. (2004)). In particular with the solution provided here the usually applied constraint of allowing only pairwise interactions of spikes could be lifted. To sum up, if we are able to successfully apply the tools derived in section 3.2 to the dynamics of synaptic connections, we could draw conclusions on the underlying plasticity mechanisms of behaving biological systems.

Temporal difference learning versus differential Hebbian plasticity

A tropism only works if the stimulus is directly perceivable, hence if it is not "delayed". If this is not the case, we would need to apply mechanisms such as secondary conditioning. We could think of a secondary conditioning as a second-order delay task where not a perceivable stimulus is rewarded but the stimulus that follows after. In detail, in secondary conditioning an already learned conditioned stimulus acts as a rewarded or unconditioned stimulus for another, yet earlier, stimulus. This was, for instance, investigated in the context of a food retrieval task in Thompson et al. (2008), however, with a modified plasticity rule. The modifications made it possible to relate the stimulus setup and the results to the structure and the function of limbic system. By contrast the normal way to learn from the delayed rewards in higher-order delay tasks is temporal difference learning. The reason for this is that the learning rule is shaped in such a way that the value converges to the return. And the return is exactly the sum of all upcoming, thus delayed, rewards if following a certain policy (for more details see introduction of chapter 4). It was used in Sutton (1988) to overcome problems of the S&B model when dealing with secondary conditioning.

In chapter 4 we described a novel setup which allowed us to emulate the temporal difference learning rule by using homosynaptic differential Hebbian plasticity. Older studies had suggested that temporal difference learning relies fundamentally on a third factor, namely the δ error (see equation 2.21). This δ error evaluates the reward by comparing it with the expectation made by the value that should represent the return (see above). Note that the reward is only represented in the third factor, and there are no reward states in the TD model. In current work the δ error is related to the Dopamine signal in the brain (see section 2.3 and Schultz et al. (1992); Schultz (1998)). However, the results presented in chapter 4 do not rely on these requirements, namely on an *evaluative* third factor and on an *implicit* reward signal. In the model developed here, the reward does not have to be a separate entity but could be just another "salient" stimulus. This stimulus could be an (unconscious) stimulus of pleasure (e.g. "This berry tastes well") or a (conscious) stimulus of will (e.g. "I wanted to move here"). Therefore, every stimulus could serve as a rewarding stimulus as long as it essentially (i.e. above noise level) drives the "learning" neuron. Second, the results in chapter 4 suggest in contradiction to TD model that a possible third factor is only modulating and not evaluating plasticity. Additionally, in section 4.2.3 we showed that it is even possible to totally omit a third factor by using different time scales for the plasticity and the output pathway. This shifts the auto-correlation contribution to non-zero values. Negative auto-correlation values are achieved when the time scale of plasticity is greater than the time scale of the output pathway. For instance, if the EPSP measured at the axon has a width of about 5 ms (Dudek and Bear, 1993), the plasticity should take at least slightly more than 5 ms (see equation 2.14 and 2.16 and Figure 4.12 for an illustration). Although the approach that both mechanisms are identically except for the time scale is questionable, the requirement for plasticity to have a duration of about 10 ms is reasonable. To evaluate this model with an experiment, we would need to manipulate the time scales of the plasticity mechanisms. If we increase the speed of plasticity, then we should find that decisions after learning are still random as the weights of all alternatives diverged or rather reached a boundary; note that weights in biological systems can not diverge due to natural limitations.

Using different time scales is only one of many possibilities to shift the auto-correlation to negative values. Note that a negative auto-correlation is a necessary condition to emulate temporal difference learning (see section 4.1), and therefore plain differential Hebbian plasticity can not be used. A prominent extension is an additional factor, the third factor (see section 2.5), which is similar to the third factor used in the current models of temporal difference learning (Schultz et al., 1992; Schultz, 1998). Furthermore, the results of Pawlak and Kerr (2008) which showed that a third factor, namely Dopamine, is necessary for long-term potentiation in the Striatum, demand that Dopamine is present while plasticity takes place. However, the third factor used in this work is different in a particular point; it is not evaluating the reward, it is only modulating plasticity (thus shifting the auto-correlation to negative values). Along the line of a non-evaluating third factor, Redgrave and Gurney pointed in Redgrave and Gurney (2006) to timing problems concerning the Dopamine signal. In particular, they showed that Dopamine, which, according to the TD model (Schultz et al., 1992), carries information about the reward, is released before the animal is conscious about the reward; however, for an unconscious evaluation of the rewarding stimuli the timing might be just right. The authors also presented an alternative interpretation of Dopamine, which is that of a novelty signal, i.e. it is elicited due to novel stimuli. This interpretation would also fit to the necessity of a decreasing learning rate in uncertain environments when using temporal difference learning. However, the properties of the third factor needed for the equivalence presented here are not conform with the properties of Dopamine because its timing in general is too imprecise. We already mentioned that a neuromodulator which better fits the properties is Acetylcholine (Graybiel, 1998). There are two starting points for possible experiments which could distinguish between the current model of temporal difference learning and the model presented in this work. If we could show that Acetylcholine or neuromodulators with similar properties are not essential for plasticity, the models presented in subsection 4.2.1 and 4.2.2 would need a different interpretation. Concerning Dopamine, an interesting experiment would be to keep the dopamine level high independently of the reward probability (it is known that the dopamine concentration varies with respect to reward probability (Fiorillo et al., 2003; Morris et al., 2004; Tobler et al., 2005)) and check whether the animal is still able to distinguish between different reward probabilities.

Another approach which does not depend on a explicit third factor was developed by Roberts et al., who showed the equivalence between asymmetrical spike-timing dependent plasticity (STDP) and temporal difference learning when having consecutive states (Roberts et al., 2009). Here, asymmetrical means that the weight-change or STDP curve has both a symmetrical and an antisymmetrical part (see Figure 5.1). In particular he showed the symmetrical part needs to be negative in order to achieve equivalence. That the framework of

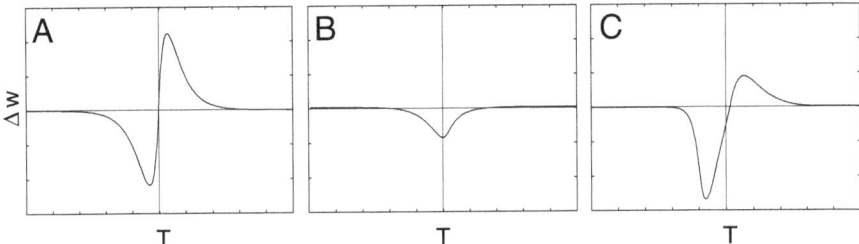

Figure 5.1: *Asymmetrical spike-timing-dependent plasticity. The weight change Δw is plotted with respect to the temporal difference of two spikes. Panel A shows the symmetrical, panel B the negative asymmetrical and panel C the summed STDP/weight change curve. The negative symmetrical weight change curve corresponds to a negative auto-correlation contribution.*

differential Hebbian plasticity and spike-timing-dependent plasticity are closely linked together was shown in chapter 2 and in work of Roberts (Roberts, 1999) himself. Thus, we conclude that the negative symmetrical part of STDP exactly corresponds to a negative auto-correlation contribution. However, the origin of this negative symmetrical part is not clear and it would be worth to pursue this experimentally.

Closed-loop extension of temporal difference learning

Behaving systems can not solely rely on temporal difference learning or the here presented emulation of it as both models are open-loop algorithms, as seen in Figure 4.7 and Figure 4.11. In these figures we set the policy, thus the actions were independent of the output or any other signal produced by the algorithm. If we want to extend temporal difference learning in such a way that the algorithm also determines its policy, we would need to add an actor-like component. For this, there exist two popular methods, namely Actor-Critic and SARSA/Q learning, and we will discuss both. Additionally we show possible extensions of our algorithm which emulates these methods.

The Actor-Critic model (Witten, 1977) consists of a separate actor module which has been added to the critic. The critic, which is the already existing temporal difference learning rule, criticizes the actor by means of the δ error (see Figure 5.2 A). As a result of this, the actor builds state-actions mappings, which provides a controller with actions leading to the reward(s). Additionally, Actor-Critic models can be related to structure and function of the basal ganglia (Joel et al., 2002). However, it is also possible to use the state-value mapping $V(x)$ of the critic to guide the agent, thus combining both modules (see Figure 5.2 B). This is because values of each state represent future rewards, and following this gradient will directly lead to the reward(s) (this is similar to klinokinesis (Spudich and Koshland, 1975) and was used to model bee foraging in Montague et al. (1995)). We also implemented a similar combined Actor-Critic (see Figure 5.3) and the results show that the performance is robust in simple environments, even when using randomly overlapping states. We were also able to map the structure of our

CHAPTER 5 DISCUSSION AND OUTLOOK

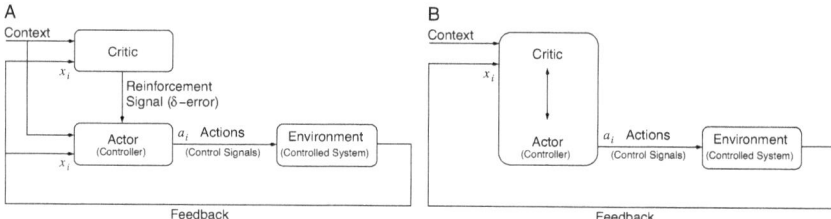

Figure 5.2: *Actor-Critic architecture. In panel A the Actor-Critic with separate actor and critic is shown. Both the actor and the critic receive states x_i and other relevant context informations. The δ error of the critic also drives the learning of the actor, which then chooses the optimal action. In panel B, we show the combined Actor-Critic model proposed here. Here the actor chooses the action which leads to a state with the highest value $V(x_{i+1})$.*

algorithm to the direct pathway of the basal ganglia (see Figure 5.3 A and Kolodziejski et al. (2008)).

The second possible extension is called Q- or SARSA learning (Watkins and Dayan, 1992; Singh et al., 2000). There, the state space is extended, namely by actions a_i an agent is able to execute at a certain state x_i. Therefore the update rule results in

$$Q(x_i, a_i) \to Q(x_i, a_i) + \alpha \left[r(x_{i+1}) + \gamma Q(x_{i+1}, b) - Q(x_i, a_i) \right] \quad (5.1)$$
$$\text{where } b = argmax_a Q(x_{i+1}, a) \quad \text{for Q learning}$$
$$\text{and } b = a_{i+1} \quad \text{for SARSA learning.}$$

The difference between Q and SARSA learning is only related to the learning rule. While in SARSA learning the Q values are updated by using the actual taken action, the Q values in Q learning are modified by taking the action which leads to the optimal policy. Thus, Q learning converges in most cases faster as the agent can explore and exploit (at least the learning values) at the same time (Sutton and Barto, 1998). As can be seen in equation 5.1, the Q values for the extended state-action space are basically still learned according to the temporal difference rule.

If we wanted to extend our setup to emulate either Q- or SARSA learning, the first step would be to include a repertoire of actions a_j (see Figure 5.4). We did this by connecting all inputs to separate dendrites d_j, where each dendrite represents an action. To this end we connected each dendrite d_j with a corresponding action a_j. In order to correlate activity at two different dendrites, all of them are additionally connected to a neuron where the weights change according to the plasticity rules presented in chapter 4. The output v of this neuron is, however, not used. Whenever a state x_i is active, it will drive the activity at a dendrite depending on the corresponding weight w_{ji}. This activity is then integrated at the action neurons a_j and as soon as a certain threshold is reached, the action corresponding to this "winning" dendrite is executed. Noise at the dendrites guarantees that the selection of the

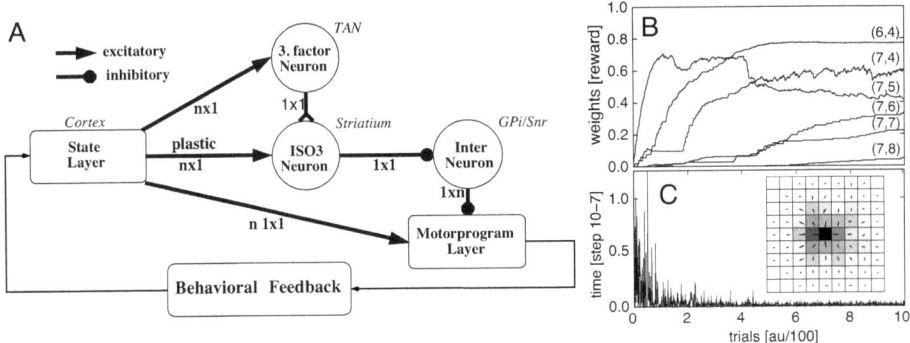

Figure 5.3: *Combined Actor-Critic model and the relation to the direct pathway of the basal ganglia. In panel A, we show the closed-loop Actor-Critic model. The tonically active neurons (TAN - third factor) modulate plasticity at neurons in the Striatum (ISO3 neuron). Their activity disinhibits the motor neurons via the globus palidus (GPi) and the substantia nigra pars reticulata (SNr). Only the excitatory connection between the state layer and the ISO3 neuron using a global third factor are plastic. Panel B and C depict the simulation results when using the Actor-Critic model in a typical grid world (inset of panel C - size 9×9). There the weights are represented by a gray shading, where black stands for 1 and white for 0. Each arrow points in the direction of the highest value. In panel B we show the time development of some selected weights (at coordinate (x,y)) and the panel C shows the time needed to reach the reward state (the black center square of the inset - coordinates: $(5,5)$). For the simulations we used $u(t) = \int_0^S (e^{-a(t-z)} - e^{-b(t-z)}) \, dz$, $a = 0.006$, $b = 0.0066$, $T = 300$, $O = 0$, $L = 90$, $\mu = 0.001$ and random initial positions. The parameters lead to a γ value of 0.73.*

actions is not only greedy, thus to allow a certain level of exploration. It shows that it is difficult to emulate either plain Q- or SARSA learning. For the latter the activity of all the other dendrites but the "winning" dendrite needs to be inhibited after the decision was made as SARSA learning requires the Q value of the actual taken action only. By contrast, to emulate Q learning, the same deciding procedure needs to take place in the absence of noise to determine the optimal action. In our preliminary implementation the weights develop according to a mix of both Q- and SARSA learning, however, yielding convergent weights and stable behavior (see Figure 5.4 B).

Relation to decision making and planning

Implementations of closed-loop algorithms like Actor-Critic or Q/SARSA learning using correlation-based learning rules allow us to bring the framework of avoidance learning and the framework of goal-directed learning together. In doing so we would be able to better explain the behavior of animals, as it is required to know which of the above strategies, punishment

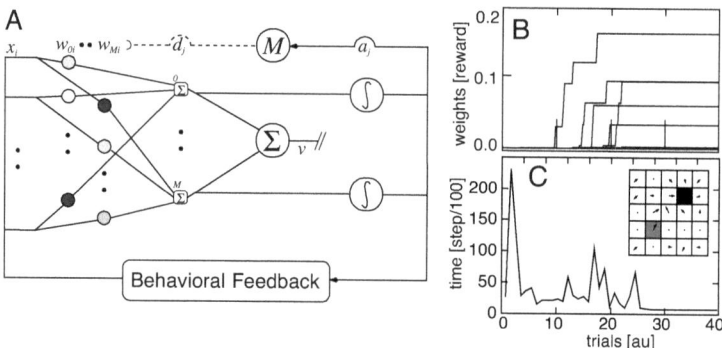

Figure 5.4: *Mixed Q/SARSA learning architecture. Panel A shows the architecture of the Q/SARSA learning architecture. There are $N \times D$ connections between the states x_i and the dendrites d_i. N and D are the number of states and dendrites, respectively. All dendrites converge to the ISO3 neuron. The plasticity is modulated by the third factor M which is triggered by the actions a_i. Panel B and C depict the simulation results when using the mixed Q/SARSA model in a typical grid world (inset of panel C - size 5×5). Each arrow points in the preferred direction given by the weight values. In panel B we show the time development of some selected weights and the panel C shows the time needed to reach the reward state (the black square of the inset - coordinates: $(4, 4)$). For the simulations we used $u(t) = \int_0^S (e^{-a(t-z)} - e^{-b(t-z)}) \, dz$, $a = 0.006$, $b = 0.0066$, $T = 300$, $O = 0$, $L = 90$, $\mu = 0.1$ and a fixed initial positions at coordinate $(2, 2)$. The parameters lead to a γ value of 0.73.*

avoidance or reward seeking, is employed in a given situation. There exist mainly two ways to solve this problem. Either, by ways of switching, the agent makes a decision for each situation between these two strategies or it uses a combination of them. An example which shows the difference between the two mentioned solutions is a behaving agent that needs to find rewards (food) in the environment (e.g. to keep their energy level high) while at the same time avoiding painful situations like obstacles or traps. It is a non-trivial problem to resolve the conflict between aversive and attractive stimuli. For example, if reward and pain lie close together the agent using the decision making policy might decide to avoid the pain, thus ending up at starvation. An agent, however, which uses a combined framework will not encounter this dilemma as the aversive stimulus does not completely inhibit the attractive action. However, this would relate differential Hebbian plasticity to the field of decision making or even planning. Nonetheless, it would be interesting to develop a combined mathematical framework for the combination policy. The resulting network could then be tested on an agent that forages in a hostile environment and would, at the end, need to be generalized such that it can be used on any given agent.

"If the human brain were so simple that we could understand it, then we would be so dumb not to understand it."

Jostein Gaarder (Sophie's World)

Appendix A
Biophysical Basics

In order to understand the biophysics of synaptic plasticity mechanisms, we will explain in this appendix the basics of activation mechanisms and how activation travels from the pre-synaptic sites to the post-synaptic site.

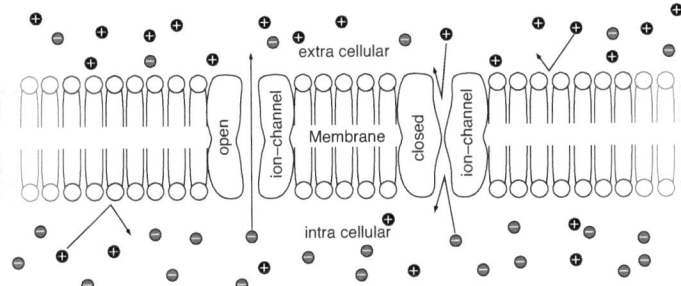

Figure A.1: *Sketch of a membrane with ion channels. The membrane is impermeable to ions which can only move between intra- and extra cellular regions by means of ion pumps (not shown) and ion channels. Depending on several influences the ion channels are either open or closed. As ions are charged particles, an electrical gradient develops. The intra cellular region usually has more negative ions and is therefore negative (about $-70\,mV$).*

Normally a cell is enveloped by a membrane, and this is also true for neurons. This membrane is impermeable to all the ions which exist in the cellular fluid. By special mechanisms, different concentrations of these ions develop in the intra and extra cellular regions (see Figure A.1). As ions are charged particles, the different concentrations cause a gradient in the potential. In more detail, there are two mechanisms that allow ions to penetrate the membrane. Ion pumps actively shift the concentration under an expense of energy from one site to an other and, more important for the understanding of activation, by means of ion channels in a passive manner. These ion channels change their configuration, thus allowing or disallowing ions to pass. These configuration changes happen for instance at different voltages or through neurotransmitters (which will be important for the transmission of the activation). If enough of them open and the concentration changes rapidly, so does the potential and an *action potential* originates (see Figure A.2 A). Whenever we talk about activation in the biophysical sense, we usually think of these action potentials or shortened, *spikes*. After origin they start to propagate (in both directions). Activity flows in a passive manner along the dendrites and/or the axon-like currents flow along a conductor (see Figure A.2 B left part). However, before the action potential reaches a synapse it might not exist anymore as dendrites and axons are not ideal conductors. Hence, a second active mechanism takes over, where the activity is transferred from one ion channel to the next. Thus the altered potential opens adjacent ion channels, i.e. a cascade of ion channel openings propagates along the neurons branches (see Figure A.2 B right part).

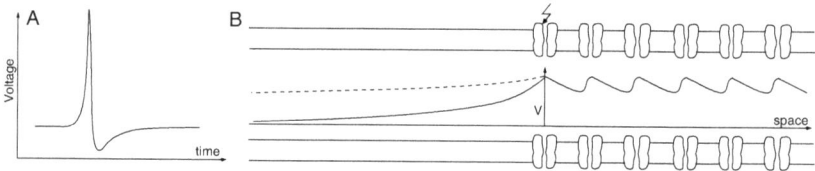

Figure A.2: *Sketch of an action potential and its propagation along an axon. Panel A shows the time development of an action potential with its steep rising phase and its slightly shallower relaxation phase. In panel B we show the mechanism of action potential propagation along an axon. An action potential was elicited at the middle ion-channel from which it propagates in a passive manner to the left and in an active way to the right. Depending on the isolation of the axon the voltage decays slowly (dashed line - good isolation) or fast (solid line) to resting potential. In order to reach a synapse the axon needs either to be sufficiently isolated or it needs adjacent open ion channels. However, the latter is more energy consuming.*

Two different kinds of synapses, electrical and chemical, exist. The first are rare and at these gap junctions, the action potential influences the post-synaptic ion channels directly, thus passing almost without a time delay. For the latter the action potential needs to trigger the release of neurotransmitter for transmission. At the pre-synaptic site usually many vesicles containing neurotransmitters exist (see Figure 1.4 upper part). When an action potential arrives, these vesicles bind to the membrane and thereby release their neurotransmitters into the synaptic cleft, which is the space between pre- and post-synaptic site. The neurotransmitters then bind to the ion channels at the post-synaptic site, which leads to a change in the potential which in turn may lead to an action potential at the post synaptic site by means of a self-energizing effect. At this point we are able to link two compartments to the synaptic efficiency. The first is the number of vesicles and their neurotransmitters and the second is the number of ion channels at the post-synaptic site.

Appendix B

Quasi-Static Weight Changes

Throughout the whole thesis we are using a quasi-static approach. For this we assume that the change in weights w_i is much smaller than the change in the convoluted signals u_i. This writes short as

$$\frac{\dot{w}(t)}{w(t)} \ll \frac{\dot{u}(t)}{u(t)}. \tag{B.1}$$

The parameter which adjusts this ratio is the plasticity rate μ (in chapter 2) or α (in chapter 4). In the following we will use μ. To this end, the plasticity rate needs to be very small or even asymptotically zero: $\mu \to 0$.

Using this quasi-static approach allows mainly for two simplifications when calculating the analytical solutions. In this appendix we will discuss these simplifications. For this we use two inputs (u_0 and u_1), similar to chapter 2, where only the weight $w_1(t)$ is time dependent. This weight changes according to $\dot{w}_1(t) = \mu\, u_1(t)\, \dot{v}(t)$, and the derivative of the output writes as $\dot{v}(t) = w_0\, \dot{u}_0(t) + w_1(t)\, \dot{u}_1(t) + \dot{w}_1(t)\, u_1(t)$. Together we have the non-simplified plasticity rule:

$$\dot{w}_1(t) = \mu\, u_1(t)\, [w_0\, \dot{u}_0(t) + w_1(t)\, \dot{u}_1(t) + \dot{w}_1(t)\, u_1(t)] \tag{B.2}$$

which we will solve in two different ways: with and without the quasi-static assumption, starting with the latter.

B.1 General solution

In general we need to solve equation B.2 using also the weight derivative of the right hand side. We bring this term to the left hand side, which results in

$$\begin{aligned}\left[1 - \mu\, u_1^2(t)\right] \dot{w}_1(t) &= \mu\, u_1(t)\, w_0\, \dot{u}_0(t) + \mu\, u_1(t)\, w_1(t)\, \dot{u}_1(t) \\ \dot{w}_1(t) &= \mu\, \frac{u_1(t)\, \dot{u}_1(t)}{1 - \mu\, u_1^2(t)}\, w_1 + \mu\, \frac{u_1(t)\, \dot{u}_0(t)}{1 - \mu\, u_1^2(t)}\, w_0 \end{aligned} \tag{B.3}$$

In order to divide by $1 - \mu u(t)^2$ we have to make sure that $\mu < u(t)^2$ for all t. Equation B.3 is an inhomogeneous differential equation of first order. The solution of this kind of equations is always the solution of the homogeneous part (auto-correlation) plus a particular solution of the inhomogeneous part (cross-correlation). The latter we get by using the method of variation of parameters, which will be discussed together with the quasi-static solution. Therefore we provide here only the homogeneous solution which we depict with the index ac to be consistent with the main text:

$$w_1^{ac}(t) = \frac{w_1(t_0)}{\sqrt{|1 - \mu\, u_1(t)^2|}} \tag{B.4}$$

where t_0 is the time the weight change starts. To compare this equation to the quasi-static solution which we find throughout this thesis we need to Taylor expand equation B.4 (neglecting the absolute value bars) around small values of μ:

$$\frac{1}{\sqrt{1 - \mu\, \xi}} = \sum_{n=0}^{\infty} \frac{(2n-1)!!\, \xi^n\, \mu^n}{2^n\, n!} = 1 + \frac{\xi\mu}{2} + \frac{3\,\xi^2\,\mu^2}{8} + o(\mu^3) \tag{B.5}$$

where we define $\eta!! = \eta\,(\eta - 2)\,(\eta - 4)\cdot\ldots\cdot 1$ for all odd values of η and $\eta!! = 1$ for all $\eta \leq 0$.

B.2 Quasi-static solution

If the weight change is small, thus the system is quasi-static, we can neglect the weight derivative on the right hand side of equation B.2, which results in

$$\dot{w}_1(t) = \mu\, u_1(t)\, w_0\, \dot{u}_0(t) + \mu\, u_1(t)\, w_1(t)\, \dot{u}_1(t) \tag{B.6}$$

The homogeneous (ac) solution of this equation is

$$w_1^{ac}(t) = w_1(t_0) \cdot \exp\left(\mu\, \frac{1}{2}\, u_1^2(t)\right) \tag{B.7}$$

A step by step calculation can be found in appendix D. We need to Taylor expand this equation, too, which results in

$$\exp\left(\mu\, \frac{1}{2}\, \xi\right) = \sum_{n=0}^{\infty} \frac{\xi^n\, \mu^n}{2^n\, n!} = 1 + \frac{\xi\mu}{2} + \frac{\xi^2\,\mu^2}{8} + o(\mu^3) \tag{B.8}$$

By comparing the Taylor expansions of the general and the quasi-static solution it becomes clear that only above the second order both equations become different. Thus, for small values of μ it is justified to use the quasi-static solution.

B.3 Variation of parameters

Next we will take a closer look at the inhomogeneous solution which corresponds to the cross-correlation contribution. To calculate the solution we need to apply the method of the variation

B.3 VARIATION OF PARAMETERS

of parameters which results in a form (using cc) which is valid for both the general and the quasi-static solution:

$$w_1^{cc}(t) = w_1^{ac}(t) \int_{t_0}^{t} \frac{w_0 \, \dot{u}_0(z) \, u_1(z)}{w_1^{ac}(z)} \, dz \qquad (B.9)$$

The Taylor expansions (equations B.5 and B.8) depict that there is a variation of the homogeneous solution from the first order on. However, as the plasticity rate μ goes to zero we can neglect the variation and pull the homogeneous solution out of the integral. This results in the equation we used throughout this thesis for the calculation of the cross-correlation contribution:

$$w_1^{cc}(t) = \int_{t_0}^{t} w_0 \, \dot{u}_0(z) \, u_1(z) \, dz \qquad (B.10)$$

It is nonetheless interesting that the weight change curve for plain differential Hebbian plasticity varies for different plasticity rates which is shown in Figure B.1. This shows that the weight change curve becomes anti-symmetrical when using large plasticity rates.

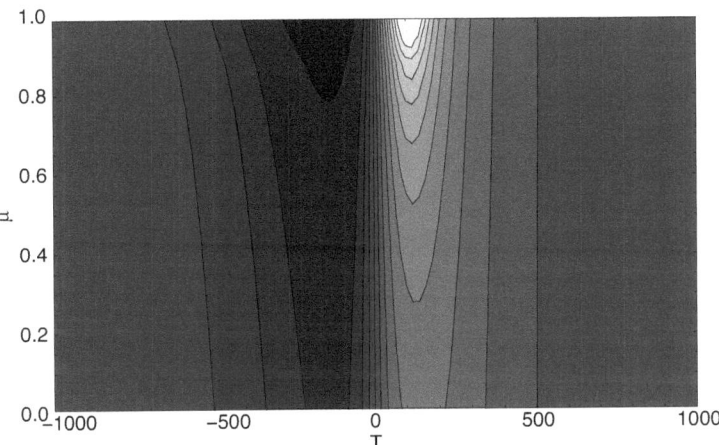

Figure B.1: *Weight change curve of differential Hebbian plasticity under different plasticity rates using the full solution. It shows that the weight change curve becomes more asymmetrical with larger plasticity rates. By contrast low plasticity rates ($\mu < 0.2$) leave the weight change curve unaffected.*

If we put both solutions together and only take the final weight after two inputs into account we arrive at

$$w_1(t) = w_1^{ac}(t) + w_1^{cc}(t) = f(t) \, w_1(t_0) + \int_{t_0}^{t} w_0 \, \dot{u}_0(z) \, u_1(z) \, dz$$

$$\Delta w_1 = (1 - f(t)) \, w_1 + \int_{t_0}^{\infty} w_0 \, \dot{u}_0(z) \, u_1(z) \, dz \qquad (B.11)$$

where $f(t)$ represents the factor given by the auto-correlation.

The last equation is used often throughout the main text and this appendix tried to cover the drawbacks of this simplification. However, *this appendix* also showed that, assuming a quasi-static process, the usage of equation B.11 is justified.

Appendix C

Numerical Considerations

When we want to solve the plasticity rules numerically, we could use many different methods which solve differential equations. However, we have a problem whenever we do not know which input will come next. This is a situation all behaving systems must face as they live in an uncertain environment. The only method which applies if you do not know the future is the basic Euler method. To this end you assume a small value which you constantly add to your differential equation. For behaving systems even another problem needs to be considered, because they usually can not choose this small value themselves. It is given by the update frequency, which is governed by the time their sensors provide the controller with new signals. The lower this update frequency is the higher are the numerical errors and the higher will be, for instance, the auto-correlation contribution in differential Hebbian plasticity although it is asymptotically converging to zero (compare section 2.1).

In robotic applications we have the possibility to change the update frequency or rather change the parameters of our kernel function. In the following we will provide an estimation of the numerical error for different parameters by comparing the integration of the kernel function equation 1.1 with the summation of it. Integrating from zero to infinity neglecting σ gives us

$$H_{\text{int}}(a,b) = \int_0^\infty h(z)\, dz = \int_0^\infty \left(e^{-at} - e^{-bt} \right) dz = \left[-\frac{1}{a} e^{-at} + \frac{1}{b} e^{-bt} \right]_0^\infty = \frac{1}{a} - \frac{1}{b} = \frac{b-a}{ab} \quad \text{(C.1)}$$

We need to compare this analytical solution C.1 with the numerical summation for which we set $t \in \mathbb{N}$ as this is the natural way to integrate the kernel numerically. This results in

$$H_{\text{sum}}(a,b) = \sum_{n=0}^\infty h(n) = \sum_{n=0}^\infty \left(e^{-an} - e^{-bn} \right) = \frac{1}{1-e^{-a}} + \frac{1}{1-e^{-b}} = \frac{e^{-a} - e^{-b}}{(1-e^{-a})(1-e^{-b})} \quad \text{(C.2)}$$

Although equations C.1 and C.2 look different, they have similar values in particular in the limit of a to zero. To depict this in a better way we set $b = 2a$ and calculate the relative difference $\Delta h(a)$

$$\Delta h(a) = \frac{H_{\text{int}}(a, 2a) - H_{\text{sum}}(a, 2a)}{H_{\text{int}}(a, 2a)} = 1 - 2a \frac{e^{-a}}{1 - e^{-2a}} = 1 - \frac{a}{\frac{1}{2}(e^a - e^{-a})} = 1 - a \sinh^{-1} a \quad \text{(C.3)}$$

APPENDIX C NUMERICAL CONSIDERATIONS

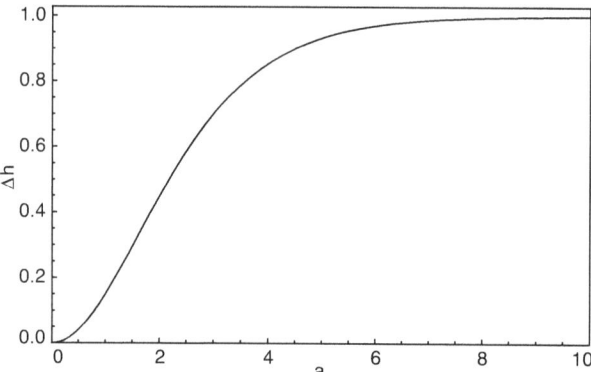

Figure C.1: *Numerical error (equation C.3) of the kernel functions $h(t)$ with respect to the parameter a. We set the other parameter $b = 2\,a$.*

The difference $\Delta h(a)$ is plotted in Figure C.1 with respect to a. The value a determines the position of the maximum, thus the length of the kernel function defined in equation 1.1. The dependence on different a values is shown in Figure 1.7 where it shows that the higher we choose a, the less space in time is covered by the kernel functions. Coming back to the numerical considerations, with a value of $a = 0.1$ the kernel function covers just about 40 time steps which are significantly about zero. Here significant means that the value decayed below 10^{-3} of the maximum value. Although only 40 time steps were used to discretize the kernel function, values of a smaller than 0.1 result, according to equation C.3, in a difference between analytical and numerical solution which seems to be tolerable. An a value of about 0.01 even decreases the error to an almost neglectable value. A comparison of the difference between $a = 0.1$ and $a = 0.01$ in a real example is plotted in the upper inset of Figure 2.3 A in the main text. The insets reveal that a discrepancy between analytical and numerical solution of about 0.1 is still large enough to induce visible numerical errors. Along this lines, it is not advisable to use a values larger than 1 and especially not larger than 5 as Figure C.1 indicates.

Appendix D

Solution of the Homogeneous Part of the General Differential Hebbian Plasticity Equation

Here we provide a solution for the homogeneous part of the general differential Hebbian equation which can be found in many equations (e.g. equation 2.5 together with equations 2.9, 2.25 or 4.4). For the following analysis we neglect the index i and include the third factor M to be as general as possible:

$$\dot{w}(t) = \frac{dw(t)}{dt} = \alpha \cdot u(t)\,\dot{u}(t)\,w(t)\,M(t) \quad \text{(D.1)}$$

Next we separate the variables and integrate both sides from zero to infinity:

$$\int_{w_0}^{w} \frac{dw}{w} = \alpha \cdot \int_0^{\infty} u(t)\,\dot{u}(t)\,M(t)\,dt \quad \text{(D.2)}$$

In the main text we mentioned that $M(t)$ is either 1 or 0, thus we can model this function as a sum of Heaviside functions $\Theta(t)$:

$$M(t) = \sum_m \Theta(t - B_m^{lower})\,\Theta(B_m^{upper} - t). \quad \text{(D.3)}$$

As the Heaviside functions determine the integration boundaries of the right hand side of equation D.2, this can be simplified to:

$$\int_{t_0}^{t} u(t)\,\dot{u}(t) \sum_m \Theta(t - B_m^{lower})\,\Theta(B_m^{upper} - t)\,dt = \sum_m \int_{B_m^{lower}}^{B_m^{upper}} u(t)\,\dot{u}(t)\,dt \quad \text{(D.4)}$$

The left side of equation D.2 solves to a logarithmic function and for its right side we use following derivative:

$$\frac{du^2(t)}{dt} = 2\,u(t)\,\dot{u}(t) \quad \text{(D.5)}$$

APPENDIX D SOLUTION OF THE HOMOGENEOUS PART OF THE GENERAL DIFFERENTIAL HEBBIAN PLASTICITY EQUATION

All in all this leads to:
$$\ln \frac{w}{w_0} = \frac{\alpha}{2} \sum_m [u^2(B_m^{upper}) - u^2(B_m^{lower})] \tag{D.6}$$

where we have to invert the logarithmic function:
$$w = w_0 \cdot \exp \frac{\alpha}{2} \sum_m [u^2(B_m^{upper}) - u^2(B_m^{lower})]. \tag{D.7}$$

This way we get the weight w after integrating over a pulse pair if having weight w_0 before.

Appendix E

Switching Integral and Derivative to Solve the Derivative of the Exponential Integral

In section 3.1 we show that the weights of differential Hebbian plasticity with a third factor and a bank of kernel functions self-organize to a situation where the auto-correlation contribution vanishes. There we encounter following integral

$$g_v(t) = \lim_{\epsilon \to 0} \int_\epsilon^\infty \frac{e^{-\xi(t)\eta}}{\eta} d\eta \tag{E.1}$$

which is a special case of the exponential integral $E_n(\xi) = \int_1^\infty e^{-\xi\eta}/\eta^n \cdot d\eta$ with $n = 1$. As we are only interested in the derivative of this integral, we do not need to solve this integral. In fact we need to solve:

$$\dot{g}_v(t) = \lim_{\epsilon \to 0} \frac{d}{dt} \int_\epsilon^\infty \frac{e^{-\xi(t)\eta}}{\eta} d\eta. \tag{E.2}$$

If $\frac{e^{-\xi(t)\eta}}{\eta}$ is continuously differentiable, then it is possible to switch the derivative and integral. As $\xi(t)$ is always a linear function in equation 3.11, the exchange is possible and we get

$$\begin{aligned}
\dot{g}_v(t) &= \lim_{\epsilon \to 0} \int_\epsilon^\infty \frac{d}{dt} \frac{e^{-\xi(t)\eta}}{\eta} d\eta = \frac{d\xi(t)}{dt} \lim_{\epsilon \to 0} \int_\epsilon^\infty e^{-\xi(t)\eta} d\eta \\
&= \frac{d,\xi(t)}{dt} \lim_{\epsilon \to 0} \left(0 - \frac{e^{-\xi(t)\epsilon}}{\xi(t)} \right) = -\frac{d\xi(t)}{dt} \frac{1}{\xi(t)} = -\frac{\dot{\xi}(t)}{\xi(t)}
\end{aligned} \tag{E.3}$$

This equation is used to simplify equation 3.11 which leads to equation 3.14.

APPENDIX E SWITCHING INTEGRAL AND DERIVATIVE TO SOLVE THE DERIVATIVE OF THE EXPONENTIAL INTEGRAL

Appendix F

Estimation of the Number of Calculations for Numerical Calculation of the Temporal Development

The following code snippet shows the relevant calculations which need to be performed for every time step. Note that the method shown here for the calculation of the pre-synaptic activity (pre) is a handy way to calculate the convolution. We get the appropriate factors (fac1, fac2) by transforming the kernel into the Z-space. We will not count the checking (if clause) for differential or plain Hebbian plasticity as this checking is not necessary.

```
//calculating the convolution for each input
for all N
   pre(t) = pre(t-2) * fac1 + pre(t-1) * fac2 + input //#calc: 4
end


//calculating the output
post(t) = 0; //#calc: 1
for all N
   post(t) = post(t) + pre(t) * weight(t) //#calc: 2
end


//checking for the type of plasticity
if Hebbian plasticity
   fpost(t) = post(t) //#calc: 0 (actually not needed)
if differential Hebbian plasticity
   fpost(t) = post(t) - post(t-1) //#calc: 1
```

APPENDIX F ESTIMATION OF THE NUMBER OF CALCULATIONS FOR NUMERICAL CALCULATION OF THE TEMPORAL DEVELOPMENT

```
//updating all weights
for all N
  weight(t) = weight(t-1) + mu * pre(t) * fpost(t) //#calc: 3
end

//total #calc: 4*N + 1 + 2*N + O(1) + 3*N = 9*N + 1(2)
```

The pseudo code provided here shows us that we need $9 \cdot N + 1$ calculations per time step to numerically calculate the weight development. For differential Hebbian plasticity we have one additional calculation step.

Appendix G

Solution of the Difference Equation Given by the Overall Weight Development

In this thesis we have often encountered a difference equation which describes the overall development of the weights investigated (see for instance equation 4.27 or section 2.1). These equations can be written in a general form as

$$x_{n+1} = (1-\alpha)\, x_n + \alpha\, y. \tag{G.1}$$

Here we want to calculate to which value equation G.1 converges. This difference equation can be solved in a simpler way formulated as a differential equation:

$$\dot{x}(t) = -\alpha\, x(t) + \alpha\, y \tag{G.2}$$

Note that it is always possible to substitute time independent factors before $x(t)$ into the plasticity rate α:

$$\dot{x}(t) = -\tilde{\alpha}\, \kappa\, x(t) + \tilde{\alpha}\, \tilde{y}$$
$$\dot{x}(t) = -\alpha\, x(t) + \alpha\, \frac{\tilde{y}}{\kappa}$$
$$\dot{x}(t) = -\alpha\, x(t) + \alpha\, y. \tag{G.3}$$

In this thesis the origin of the time independent factor κ is the auto-correlation contribution Δw^{ac} or rather its negative value.

APPENDIX G SOLUTION OF THE DIFFERENCE EQUATION GIVEN BY THE OVERALL WEIGHT DEVELOPMENT

The homogeneous part solves to an exponential function with exponent $-\alpha\,t$ and the inhomogeneous solution can be found by the method of variation of parameters:

$$x_{inhom}(t) = x_{hom}(t) \int_0^\infty x_{hom}^{-1}(z)\,\alpha\,y\,dz$$

$$x_{inhom}(t) = \exp(-\alpha\,t) \int_0^t \exp(\alpha\,z)\,\alpha\,y\,dz$$

$$x_{inhom}(t) = \exp(-\alpha\,t)\,\alpha\,y \left[\frac{1}{\alpha}\exp(\alpha\,z)\right]_0^t = y\,(1 - \exp(-\alpha\,t)) \tag{G.4}$$

This gives us for the convergence:

$$x(t) = (C - y)\exp(-\alpha\,t) + y$$

$$\lim_{t\to\infty} x(t) = y \tag{G.5}$$

where C is a constant. Note that equation G.4 only converges to y if α is positive. If we relate α to the main text, it consists of the plasticity rate $\tilde{\alpha}$ and the negative value of the auto-correlation which is defined as κ. As the plasticity rate is usually positive, convergence depends on a positive κ value, hence on a negative auto-correlation Δw^{ac}. In this case difference equations like equation G.1 or differential equations like equation G.2 always converge to y. Additionally, Kushner and Clark (1978) showed that this holds also for a stochastic variable with mean y where the variance is being reduced by a time-dependent plasticity rate. To this end the plasticity rate has to decrease over time proportional to a function $f(t)$ with following properties: $\sum_{t=0}^\infty f(t) = \infty$ and $\sum_{t=0}^\infty f^2(t) < \infty$. An example function would be $f(t) = 1/t$.

In order to solve equation 4.25 we need to find in a second step the solution of another difference equation. We get this equation by assuming that equation 4.25, which is of the same form as equation G.1, already converged. This gives us following equation

$$w_n = \varepsilon_1\,w_{n+1} - \varepsilon_2\,w_{n-1}. \tag{G.6}$$

We get the solution using the Ansatz $w_n = \lambda^n$:

$$\lambda^n = \varepsilon_1\,\lambda^n\,\lambda - \varepsilon_2\,\lambda^n\,\lambda^{-1} \tag{G.7}$$

Assuming $\lambda \neq 0$ we can divide equation G.7 by λ^{n-1} and have to solve the following quadratic equation:

$$0 = \lambda^2 - \frac{1}{\varepsilon_1}\lambda - \frac{\varepsilon_2}{\varepsilon_1}$$

$$\lambda_{1/2} = \frac{1}{2\,\varepsilon_1} \pm \sqrt{\frac{1}{(2\,\varepsilon_1)^2} + \frac{\varepsilon_2}{\varepsilon_1}} \tag{G.8}$$

A negative solution, however, would lead to an oscillation, therefore we only have to consider the positive sign, which leads to a positive solution:

$$\lambda = \frac{1}{2\,\varepsilon_1} + \sqrt{\frac{1}{(2\,\varepsilon_1)^2} + \frac{\varepsilon_2}{\varepsilon_1}} \qquad (G.9)$$

that, potentiated by n, gives us the value which w_n will converge to. Therefore we can set $w_n = \lambda^{-1} w_{n+1}$ and equation G.6 is simplified.

APPENDIX G SOLUTION OF THE DIFFERENCE EQUATION GIVEN BY THE OVERALL WEIGHT DEVELOPMENT

Appendix H

Analytical Calculation of γ Using First and Second Order Terms

In this appendix we want to analytically calculate or integrate equations 4.17, 4.22 and 4.21. To this end we define a signal function guided by biophysical considerations. This function is divided into a rising phase, a plateau, and a falling phase. We expand both the rising and the falling phase to the second order. This allows us to switch between linear and quadratic rising and falling phases. After the calculation we summarize the results and, most importantly, extract information about the essential areas, i.e. areas in which the system diverges ($\kappa \leq 0$) or in which the weights of systems do not change at all ($\tau = 0$). These areas are then plotted and compared with the results obtained in chapter 4. We find that the calculations made in this appendix are transferable to other signal shapes that follow the same basic biophysical ideas.

H.1 Taylor expansion of the kernel function

The Taylor expansion to the second order of an arbitrary kernel function with a plateau is described as:

$$u(t) = U \cdot \begin{cases} 0 & \text{if } t < 0 \\ (1-\eta)\left(\frac{t}{P_E}\right)^2 + \eta \frac{t}{P_E} & \text{if } t \geq 0 \cap t \leq P_E \\ 1 & \text{if } t > P_E \cap t \leq S \\ 1 - (1-\xi)\left(\frac{t-S}{P_F}\right)^2 - \xi \frac{t-S}{P_F} & \text{if } t > S \cap t \leq S + P_F \\ 0 & \text{if } t > S + P_F \end{cases} \quad \text{(H.1)}$$

where U is the height of the plateau, P_E and P_F the length of the rising and the falling phase respectively, S the length of a state and η and ξ the degree of the second order term for the rising and the falling phase respectively. Here $\eta = 1$ or $\xi = 1$ leads to a linear slope, $\eta = 0$ or $\xi = 2$ to a convex and $\eta = 2$ or $\xi = 0$ to a concave slope.

APPENDIX H ANALYTICAL CALCULATION OF γ USING FIRST AND SECOND ORDER TERMS

I_G	$O < 0 \cap O+L < 0$	A_G	$O+T < S \cap O+L+T < S$
II_G	$O < 0 \cap O+L \geq 0 \cap O+L < P_E$	B_G	$O+T < S \cap O+L+T \geq S \cap O+L+T < P_F$
III_G	$O < 0 \cap O+L \geq P_E$	C_G	$O+T < S \cap O+L+T \geq P_F$
IV_G	$O \geq 0 \cap O < P_E \cap O+L < P_E$	D_G	$O+T \geq S \cap O+T < P_F \cap O+L+T < P_F$
V_G	$O \geq 0 \cap O < P_E \cap O+L \geq P_E$	E_G	$O+T \geq S \cap O+T < P_F \cap O+L+T \geq P_F$
VI_G	$O \geq P_E \cap O+L \geq P_E$	F_G	$O+T \geq P_F \cap O+L+T \geq P_F$

Table H.1: *The intervals used to discriminate between the different occurrence of the global third factor. See Figure H.1 for a more intuitive representation.*

I_L	$0 < T \cap O+L < T$	A_L	$0 < S \cap O+L < S$
II_L	$0 < T \cap O+L \geq T \cap O+L < P_E+T$	B_L	$0 < S \cap O+L \geq S \cap O+L < P_F$
III_L	$0 < T \cap O+L \geq P_E+T$	C_L	$0 < S \cap O+L \geq P_F$
IV_L	$0 \geq T \cap O < P_E+T \cap O+L < P_E+T$	D_L	$0 \geq S \cap O < P_F \cap O+L < P_F$
V_L	$0 \geq T \cap O < P_E+T \cap O+L \geq P_E+T$	E_L	$0 \geq S \cap O < P_F \cap O+L \geq P_F$
VI_L	$0 \geq P_E+T \cap O+L \geq P_E+T$	F_L	$0 \geq P_F \cap O+L \geq P_F$

Table H.2: *The intervals used to discriminate between the different occurrence of the local third factor are similar to the ones used for the global third factor. More precisely, we need to shift the right hand side of each inequality by T. Thus, also Figure H.1 can be used for a more intuitive representation as this figure does not depend on T.*

H.2 Intervals given a third factor

Having defined the actual shape of the kernel function, we now have to distinguish between different occurrence times of the third factor. Figure H.1 shows the six essential regions of both the rising and the falling phase. For this we defined intervals I to VI for the rising phase and A to F for the falling phase (see tables H.1 and H.2, where we defined the intervals explicitly for the global - subscript G - and the local - subscript L - third factor).

Additionally we need to define another four intervals used for the correlation of the signals when calculating $\tau_{G/L}^{\pm}$ (see table H.3).

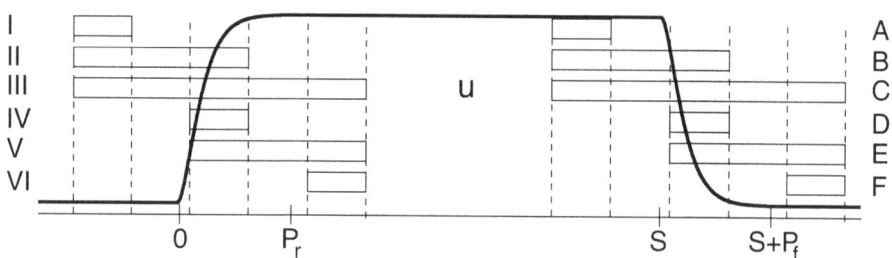

Figure H.1: *Here we show the kernel functions with all possible regions for both the rising and the falling phase. The relevant intervals are defined by I to VI and A to F.*

H.3 ANALYTICAL CALCULATION OF κ_G AND κ_L

a	$-T < 0$
b	$-T + P_F < 0$
c	$-T < P_E$
d	$-T + P_F < P_E$

Table H.3: *Intervals needed for the correlation of two consecutive signals with a time delay of T.*

Interval	$\kappa_G^+/(\frac{U^2}{2})$	Interval	$\kappa_G^-/(\frac{U^2}{2})$
I_G	0	A_G	0
II_G	$\Phi(O+L)$	B_G	$\Psi(O+S+T+L) - 1$
III_G	1	C_G	-1
IV_G	$\Phi(O+L) - \Phi(O)$	D_G	$\Psi(O+S+T+L) - \Psi(O+S+T)$
V_G	$1 - \Phi(O)$	E_G	$-\Psi(O+S+T)$
VI_G	0	F_G	0

Table H.4: *Analytical result of $\kappa_G^\pm$*

H.3 Analytical calculation of κ_G and κ_L

Next we calculate κ using the Taylor expanded function (equation H.1). For the following we assume that $O + L < S$ and $O + S + T > P_E$ which holds if S is sufficiently larger than O, L and $|T|$. These assumptions prevent cases where the third factor would effect the signal after the next signal which is nonsensical. For the global factor we simplify κ_G by splitting it into κ_G^+ and κ_G^- as done in the main text: $\kappa_G = -(\kappa_G^+ + \kappa_G^-)$. For the local third factor we set $\kappa_L = -\kappa_L^-$ as κ_L^+ does not exist. In table H.4 the analytical results for $\kappa_G^\pm$ and in table H.5 the analytical results for κ_L are stated using equations H.2 and H.3. These equations result from equation 4.30 where we included the Taylor expanded function (equation H.1). In detail equation H.2 represents the squared rising phase and equation H.3 the squared falling phase. It is important to mention that both functions, $\Phi(t)$ and $\Psi(t)$ are bounded between 0 and 1.

$$\Phi(t) = \left(\frac{t}{P_E}\right)^2 \cdot \left(\eta + (1-\eta)\frac{t}{P_E}\right)^2 \tag{H.2}$$

$$0 \leq \Phi \leq 1 \quad \forall \eta : 0 \leq \eta \leq 2 \cap \forall t : 0 \leq t \leq P_E$$

$$\Psi(t) = \left(1 - \frac{t-S}{P_F}\right)^2 \cdot \left(1 + (1-\xi)\frac{t-S}{P_F}\right)^2 \tag{H.3}$$

$$0 \leq \Psi \leq 1 \quad \forall \xi : 0 \leq \xi \leq 2 \cap \forall t : S \leq t \leq S + P_F$$

Discussion of $\kappa^\pm$: The results can be summarized by plotting the areas which represent definitive divergent areas where we simplify the rising and falling time to identical values: $P = P_E = P_F$. These areas are composed from intervals in which the sum of κ_G^+ and κ_G^- is always positive. For instance for interval A_G conjoined with either of the intervals I_G to VI_G,

Interval	$\kappa_L/(\frac{U^2}{2})$
A_L	0
B_L	$1 - \Psi(O+S+L)$
C_L	1
D_L	$\Psi(O+S) - \Psi(O+S+L)$
E_L	$\Psi(O+S)$
F_L	0

Table H.5: *Analytical result of κ_L*

the sum of κ_G^+ and κ_G^- is always greater than zero and thus divergent. The same holds for interval F_G. Both areas are indicated in Figure H.2. There is an additional divergent area; however, III_G only exists there if the value of L is greater than P.

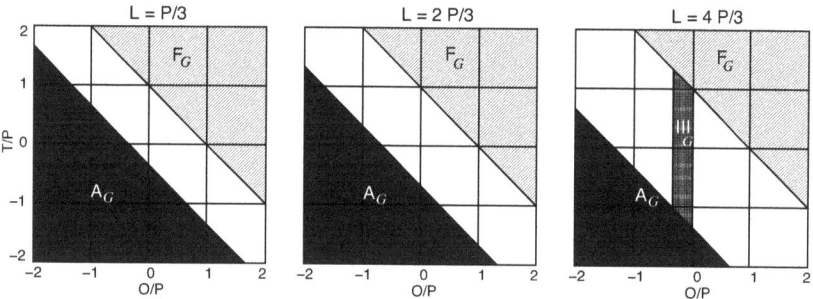

Figure H.2: *-κ_G- Here we show the divergent regions for different values of L. The intervals indicated by different patterns can be found in table H.1. The striped pattern represents regions which are independent of L, whereas the filled regions depend on L. The checked region is also depending on L, however, this area is only there for $L > P$.*

Additionally there would be also an interval C_G (not shown) for which the sum of κ_G^+ and κ_G^- is always less than zero if conjoined with interval I_G to VI_G except III_G. However, this shows up only if L is greater than P, but then interval III_G becomes valid. This can be resolved by using different values for P_E and P_F where $P_F < P_E$.

Discussion of κ_L: As $u(t)$ and with it $\Psi(t)$ are monotonically decreasing functions the only regions in which κ_L is equal to zero are A_L and F_L. Identically to the global third factor, the results can be summarized in Figure H.3, where definitive divergent areas are represented by different colors. Again, for simplicity, we take the rising and falling time to have identical values: $P = P_E = P_F$. This is not necessary for κ_L, but later for τ_L.

Only in the white area there can be convergence, however, the shape of the actual kernel function u determines whether a certain area converges or not. Additionally, it is important to

H.4 ANALYTICAL CALCULATION OF $\tau_G^\pm$ AND τ_L

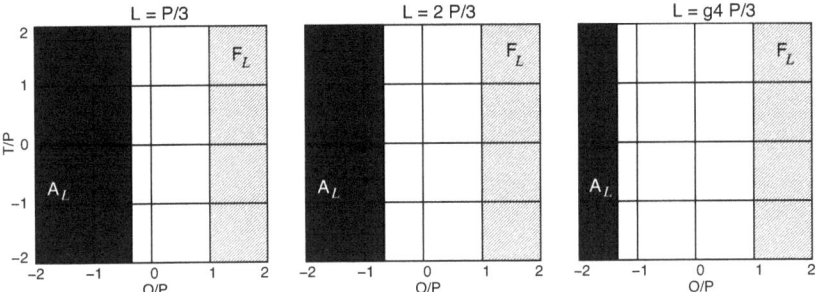

Figure H.3: -κ_L- Here we show regions of divergent γ_L values, i.e. regions in which κ_L is zero, for different values of L. The intervals indicated by the different patterns can be found in table H.2. Filled represents regions which are independent of L, whereas the striped regions depend on L.

include the values for $\tau_{G/L}^\pm$ into the considerations as these values can be 0 in different areas, which means that there is no overlap between two consecutive signals and the third factor. This will be investigated next.

H.4 Analytical calculation of $\tau_G^\pm$ and τ_L

Here we calculate the values for $\tau_{G/L}^\pm$ using the Taylor expanded function (equation H.1). We likewise assume that there is no overlap of the signal with the signal after its next state. In tables H.6 and H.7 the analytical results for $\tau_G^\pm$ and for τ_L are stated using equations H.4, H.5, H.6 and H.7. These equations result from equations 4.31 and 4.32 where we

APPENDIX H ANALYTICAL CALCULATION OF γ USING FIRST AND SECOND ORDER TERMS

included the Taylor expanded function (equation H.1). It is important to mention that these functions, $\zeta(t)$, $\psi(t)$, $\chi(t)$ and $\varphi(t)$ are always greater than 0.

$$\zeta(t) = \frac{2}{U^2} \int U \dot{u}(t)\, dt = 2\left((1-\eta)\left(\frac{t}{P_E}\right)^2 + \eta \frac{t}{P_E}\right) \tag{H.4}$$

$$\zeta \geq 0 \qquad \forall \eta : 0 \leq \eta \leq 2$$

$$\begin{aligned}
\psi(t) =& \frac{2}{U^2} \int \dot{u}(t)\, u(t+S+T)\, dt \\
=& 2\left((1-\eta)\left(\frac{t}{P_E}\right)^2 + \eta \frac{t}{P_E}\right) \\
& - (1-\eta)(1-\xi)\left(\left(\frac{t}{P_F}\right)^2\left(\frac{t}{P_E}\right)^2 + \frac{8}{3}\left(\frac{t}{P_F}\right)^2 \frac{t}{P_E}\frac{T}{P_E} + 2\left(\frac{t}{P_F}\right)^2\left(\frac{T}{P_E}\right)^2\right) \\
& - \eta(1-\xi)\left(\frac{2}{3}\left(\frac{t}{P_F}\right)^2 \frac{t}{P_E} + 2\left(\frac{t}{P_F}\right)^2 \frac{T}{P_E} + 2\left(\frac{T}{P_F}\right)^2 \frac{t}{P_E}\right) \\
& - (1-\eta)\xi\left(\frac{4}{3}\frac{t}{P_F}\left(\frac{t}{P_E}\right)^2 + 2\frac{T}{P_F}\left(\frac{t}{P_E}\right)^2\right) \\
& - \eta\xi\left(\frac{t}{P_F}\frac{t}{P_E} + 2\frac{t}{P_F}\frac{T}{P_E}\right)
\end{aligned} \tag{H.5}$$

$$\psi \geq 0 \qquad \forall \eta : 0 \leq \eta \leq 2 \cap \forall \xi : 0 \leq \xi \leq 2$$

$$\chi(t) = \frac{2}{U^2} \int U \dot{u}(t+S+T)\, dt = -2\left((1-\xi)\left(\left(\frac{t}{P_F}\right)^2 + 2\frac{t}{P_F}\frac{T}{P_F}\right) + \xi \frac{t}{P_F}\right) \tag{H.6}$$

$$\chi \leq 0 \qquad \forall \xi : 0 \leq \xi \leq 2$$

$$\begin{aligned}
\varphi(t) =& \frac{2}{U^2} \int u(t)\, \dot{u}(t+S+T)\, dt \\
=& -(1-\eta)(1-\xi)\left(\left(\frac{t}{P_F}\right)^2\left(\frac{t}{P_E}\right)^2 + \frac{4}{3}\left(\frac{t}{P_F}\right)^2 \frac{t}{P_E}\frac{T}{P_E}\right) \\
& - \eta(1-\xi)\left(\frac{4}{3}\left(\frac{t}{P_F}\right)^2 \frac{t}{P_E} + 2\left(\frac{t}{P_F}\right)^2 \frac{T}{P_E}\right) \\
& - (1-\eta)\xi\left(\frac{1}{3}\frac{t}{P_F}\left(\frac{t}{P_E}\right)^2\right) \\
& - \eta\xi\left(\frac{t}{P_F}\frac{t}{P_E}\right)
\end{aligned} \tag{H.7}$$

$$\varphi \leq 0 \qquad \forall \eta : 0 \leq \eta \leq 2 \cap \forall \xi : 0 \leq \xi \leq 2$$

H.4 ANALYTICAL CALCULATION OF $\tau_G^{\pm}$ AND τ_L

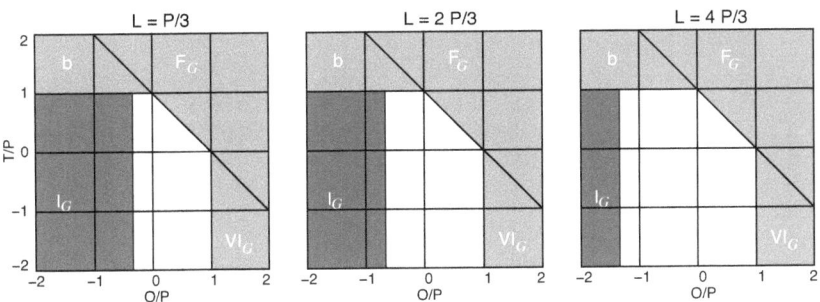

Figure H.4: τ_G^+- Here we show the regions of zero τ_G^+ values for different values of L. The intervals indicated by the different shades of gray can be found in table H.1. Light gray represents regions which are independent of L, whereas the darker gray regions depend on L.

Discussion of $\tau_G^{\pm}$: For these calculations we also simplify the rising and falling time to have identical values: $P = P_E = P_F$. The results can be summarized also by plotting the areas for which $\tau_G^{\pm}$ are zero (Figure H.4 and Figure H.5). For instance, similar to κ_G, interval F_G conjoined with intervals I_G to VI_G results in both τ_G^+ and τ_G^- to be equal to zero. In case of τ_G^- this observation can also be made for interval A_G, and in case of τ_G^+ the interval VI_G gives us zero. There are two additional intervals, I_G and b, which result in a value of zero for both τ_G measures. These areas are indicated with different colors in Figure H.4 for τ_G^+ and in Figure H.5 for τ_G^-.

APPENDIX H ANALYTICAL CALCULATION OF γ USING FIRST AND SECOND ORDER TERMS

Interval	$\tau_G^+/(\frac{U^2}{2})$ and $\tau_L/(\frac{U^2}{2})$	$-\tau_G^-/(\frac{U^2}{2})$	γ_G	γ_L
I	0	0	0	0
II $\cap$ A	$\zeta(O+L) - \zeta(0)$	0	∞	∞
II $\cap$ B $\cap$ a	$\psi(O+L) - \psi(0)$	$\varphi(O+L) - \varphi(0)$	≥ 0	≥ 0
II $\cap$ B $\cap$ $\bar{a}$	$\zeta(-T) - \zeta(0) + \psi(O+L) - \psi(-T)$	$\varphi(O+L) - \varphi(-T)$	≥ 0	≥ 0
II $\cap$ C $\cap$ a $\cap$ b	0	0	0	0
II $\cap$ C $\cap$ a $\cap$ $\bar{b}$	$\psi(-T + P_F) - \psi(0)$	$\varphi(-T + P_F) - \varphi(0)$	≥ 0	≥ 0
II $\cap$ C $\cap$ $\bar{a}$	$\zeta(-T) - \zeta(0) + \psi(-T + P_F) - \psi(-T)$	$\varphi(-T + P_F) - \varphi(-T)$	≥ 0	≥ 0
II $\cap$ D	$\psi(O+L) - \psi(O)$	$\varphi(O+L) - \varphi(O)$	≥ 0	≥ 0
II $\cap$ E $\cap$ b	0	0	0	0
II $\cap$ E $\cap$ $\bar{b}$	$\psi(-T + P_F) - \psi(0)$	$\varphi(-T + P_F) - \varphi(0)$	≥ 0	≥ 0
II $\cap$ F	0	0	0	0
III $\cap$ A	$\zeta(P_E) - \zeta(0)$	$\varphi(P_E) - \varphi(0) + \chi(O+L) - \chi(P_E)$	∞	≥ 0
III $\cap$ B $\cap$ a	$\psi(P_E) - \psi(0)$	$\varphi(P_E) - \varphi(-T) + \chi(O+L) - \chi(P_E)$	∞	≥ 0
III $\cap$ B $\cap$ $\bar{a}$ $\cap$ c	$\zeta(-T) - \zeta(0) + \psi(P_E) - \psi(-T)$	$\chi(O+L) - \chi(P_E)$	∞	≥ 0
III $\cap$ B $\cap$ $\bar{c}$	$\zeta(P_E) - \zeta(0)$	0	0	0
III $\cap$ C $\cap$ b	0	$\varphi(-T + P_F) - \varphi(0)$	∞	≥ 0
III $\cap$ C $\cap$ a $\cap$ $\bar{b}$ $\cap$ d	$\psi(-T + P_F) - \psi(0)$	$\varphi(P_E) - \varphi(0) + \chi(-T + P_F) - \chi(P_E)$	∞	≥ 0
III $\cap$ C $\cap$ a $\cap$ $\bar{d}$	$\psi(P_E) - \psi(0)$	$\varphi(-T + P_F) - \varphi(-T)$	∞	≥ 0
III $\cap$ C $\cap$ $\bar{a}$ $\cap$ d	$\zeta(-T) - \zeta(0) + \psi(-T + P_F) - \psi(-T)$	$\varphi(P_E) - \varphi(-T) + \chi(-T + P_F) - \chi(P_E)$	∞	≥ 0
III $\cap$ C $\cap$ $\bar{a}$ $\cap$ $\bar{d}$	$\zeta(-T) - \zeta(0) + \psi(P_E) - \psi(-T)$	$\chi(-T + P_F) - \chi(P_E)$	∞	≥ 0
III $\cap$ C $\cap$ $\bar{c}$	$\zeta(P_E) - \zeta(0)$	0	0	0
III $\cap$ D	$\psi(P_E) - \psi(0)$	$\varphi(P_E) - \varphi(0) + \chi(O+L) - \chi(P_E)$	∞	≥ 0
III $\cap$ E $\cap$ b	0	0	0	0
III $\cap$ E $\cap$ $\bar{b}$ $\cap$ d	$\psi(-T + P_F) - \psi(0)$	$\varphi(-T + P_F) - \varphi(0)$	∞	≥ 0
III $\cap$ E $\cap$ $\bar{d}$	$\psi(P_E) - \psi(0)$	$\varphi(P_E) - \varphi(0) + \chi(-T + P_F) - \chi(P_E)$	∞	≥ 0
III $\cap$ F	0	0	0	0

Table H.6: *Analytical results of $\tau_{G/L}^{\pm}$ and $\gamma_{G/L}$ (part I). Instead of I-VI and A-F we used the appropriate intervals for either the global ($\tau_G^{\pm}$ and γ_G: I_G-VI_G and A_G-F_G) or local (τ_L and γ_L: I_L-VI_L and A_L-F_L) third factor.*

H.4 ANALYTICAL CALCULATION OF $\tau_G^\pm$ AND τ_L

Interval	$\tau_G^+/(\frac{U^2}{2})$ and $\tau_L/(\frac{U^2}{2})$	$-\tau_G^-/(\frac{U^2}{2})$	γ_G	γ_L
IV ∩ A	$\zeta(O+L) - \zeta(O)$	0	∞	∞
IV ∩ B	$\zeta(-T) - \zeta(O) + \psi(O+L) - \psi(-T)$	$\varphi(O+L) - \varphi(-T)$	≥ 0	≥ 0
IV ∩ C	$\zeta(-T) - \zeta(O) + \psi(-T+P_F) - \psi(-T)$	$\varphi(-T+P_F) - \varphi(-T)$	≥ 0	≥ 0
IV ∩ D	$\psi(O+L) - \psi(O)$	$\varphi(O+L) - \varphi(O)$	≥ 0	≥ 0
IV ∩ E	$\psi(-T+P_F) - \psi(O)$	$\varphi(-T+P_F) - \varphi(O)$	≥ 0	≥ 0
IV ∩ F	0	0	0	0
V ∩ A	$\zeta(P_E) - \zeta(O)$	0	∞	∞
V ∩ B ∩ c	$\zeta(-T) - \zeta(O) + \psi(P_E) - \psi(-T)$	$\varphi(P_E) - \varphi(-T) + \chi(O+L) - \chi(P_E)$	≥ 0	≥ 0
V ∩ B ∩ $\bar{c}$	$\zeta(P_E) - \zeta(O)$	$\chi(O+L) - \chi(-T)$	≥ 0	≥ 0
V ∩ C ∩ d	$\zeta(-T) - \zeta(O) + \psi(-T+P_E) - \psi(-T)$	$\varphi(-T+P_F) - \varphi(-T)$	≥ 0	≥ 0
V ∩ C ∩ $\bar{d}$	$\zeta(-T) - \zeta(O) + \psi(P_E) - \psi(-T)$	$\varphi(P_E) - \varphi(-T) + \chi(-T+P_F) - \chi(P_E)$	≥ 0	≥ 0
V ∩ C ∩ $\bar{e}$	0	$\chi(-T+P_F) - \chi(-T)$	≥ 0	≥ 0
V ∩ D	$\psi(P_E) - \psi(O)$	$\varphi(P_E) - \varphi(O) + \chi(O+L) - \chi(P_E)$	≥ 0	≥ 0
V ∩ E ∩ d	$\psi(-T+P_F) - \psi(O)$	$\varphi(-T+P_F) - \varphi(O)$	≥ 0	≥ 0
V ∩ E ∩ $\bar{d}$	$\psi(P_E) - \psi(O)$	$\varphi(P_E) - \varphi(O)$	≥ 0	≥ 0
V ∩ F	0	0	0	0
VI ∩ A	0	0	0	0
VI ∩ B	0	$\chi(O+L) - \chi(-T)$	0	0
VI ∩ C	0	$\chi(-T+P_F) - \chi(-T)$	0	0
VI ∩ D	0	$\chi(O+L) - \chi(O)$	0	0
VI ∩ E	0	$\chi(-T+P_F) - \chi(O)$	0	0
VI ∩ F	0	0	0	0

Table H.7: *Analytical results of $\tau_{G/L}^\pm$ and $\gamma_{G/L}$ (part II). Instead of I-VI and A-F we used the appropriate intervals for either the global ($\tau_G^\pm$ and γ_G: I_G-VI_G and A_G-F_G) or local (τ_L and γ_L: I_L-VI_L and A_L-F_L) third factor.*

APPENDIX H ANALYTICAL CALCULATION OF γ USING FIRST AND SECOND ORDER TERMS

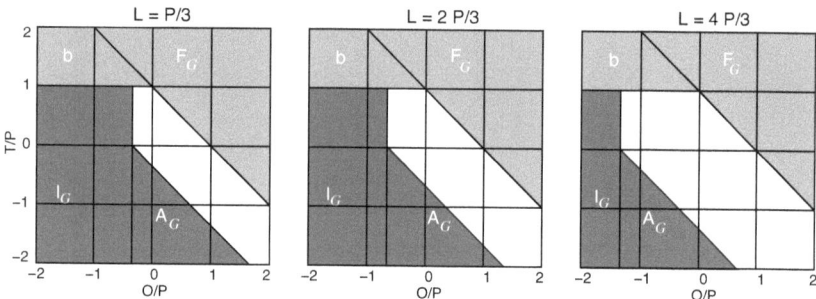

Figure H.5: τ_G^- - Here we show the regions which result in zero τ_G^- values for different values of L. The intervals indicated by the different shades of gray can be found in table H.1. For the color code see Figure H.4.

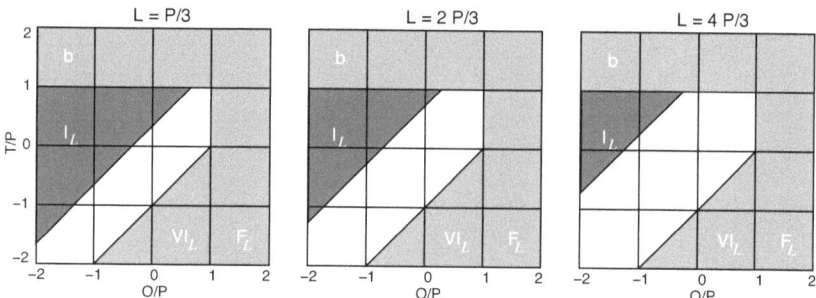

Figure H.6: τ_L - Here we show the regions of zero τ_L values for different values of L. The intervals indicated by the different shades of gray can be found in table H.1. For the color code see Figure H.4.

Discussion of τ_L: Similar to the previous paragraphs, we set $P = P_E = P_F$ and plot the results or rather areas for which τ_L is zero (Figure H.6). We find that, identical to τ_G^+, the τ_L value is zero for regions I_L, VI_L, F_L and b. However, due to the additional T-shift all regions except region b are rotated by $\pi/4$.

As $\tau_{G/L}^\pm$ will not effect convergence, all areas would yield convergence. However, only for the white area $\gamma_{G/L}$ results in a value which is unequal to zero.

H.5 Analytical calculation of γ_G and γ_L

Finally, we can calculate the value of $\gamma_{G/L}$ using equation 4.26. This is not done explicitly here, however, we indicate in the last two columns of tables H.6 and H.7 whether $\gamma_{G/L}$ is zero (no overlap between two consecutive signals and the third factor), greater than zero or infinite.

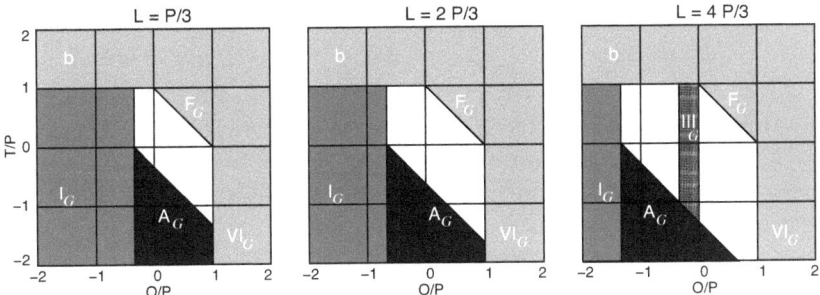

Figure H.7: γ_G- Here we show the regions of meaningless γ_G values, where there is either no overlap between two consecutive signals and the third factor ($\tau_G^\pm = 0$), or γ_G diverges ($\kappa_G < 0$). The areas are shown for different values of L. The intervals indicated by the different shades of gray and patterns can be found in table H.1. For the color code see Figure H.2 and Figure H.4. It is now possible to compare this figure with Figure 4.5 and Figure 4.6.

Furthermore, we can combine the considerations made for $\kappa_{G/L}$ and $\tau_{G/L}^\pm$ which are illustrated in Figures H.2 - H.6 into Figure H.7 for the global and into Figure H.8 for the local third factor.

If we compare Figure H.7 with Figure 4.5 and Figure 4.6 and Figure H.8 with Figure 4.9 and Figure 4.10, we find that the areas of divergence map exactly and that convergence can be found only in the white areas. Therefore the considerations about convergence and non-zero $\gamma_{G/L}$ values can be transferred from the Taylor expanded function (equation H.1) to all possible functions that possess only one plateau.

H.6 Analytical calculation of κ_T, $\tau_T^\pm$ and γ_T

In order to calculate κ_T, $\tau_T^\pm$ and finally γ_T, we need to define the rising and the falling phase of the output pathway to be different from the plasticity phase. To this end we need control parameters ρ_E and ρ_F respectively. These are set to be greater 1 as the phases for the output

APPENDIX H ANALYTICAL CALCULATION OF γ USING FIRST AND SECOND ORDER TERMS

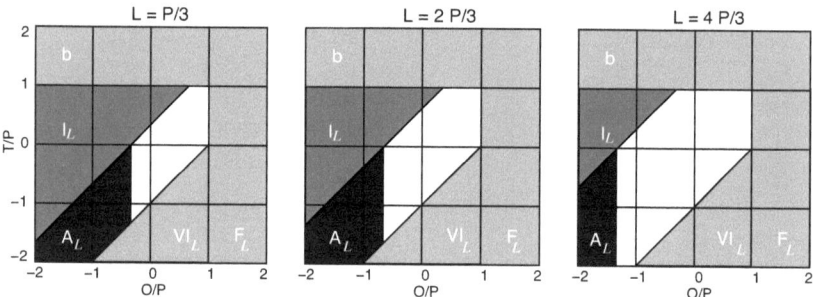

Figure H.8: γ_L - Here we show the regions of nonsensical γ values, where there is either no overlap between two consecutive signals and the third factor $(\tau^{\pm} = 0)$, or γ diverges $(\kappa = 0)$. The areas are shown for different values of L. The intervals indicated by the different shades of gray and patterns can be found in table H.1. For the color code see Figure H.2 and Figure H.4. It is now possible to compare this figure with Figure 4.5 and Figure 4.6.

pathway must be shorter. Thus, the duration of the rising phase of the output pathway is P_E/ρ_E and the duration of the falling phase P_F/ρ_F. This leads to

$$\kappa_T = -(\kappa_T^+ + \kappa_T^-) = \\ -\frac{U^2}{2}\left(\frac{3-(1-\rho_E)(4-\eta)\,\eta}{3\,\rho_E^2} + \frac{3-6\,\rho_F^2-(1-\rho_F)(4-\xi)\,\xi}{3\,\rho_F^2}\right) \quad \text{(H.8)}$$

and for the τ_T values to

$$\tau_T^- = -\frac{U^2}{2}\frac{\left(\frac{P_E}{\rho_F}-T\right)^2}{3\,P_F^2\,P_E^2}\,\Theta(\frac{P_F}{\rho_F}-T)\times \\ \left(\rho_F\,P_F\,(2\,T(1-\eta) - P_E\,\eta\,(4-\xi)) - P_F^2\,(1-\eta)(3-\xi)\right. \\ \left. + \rho_F^2\,T\,(1-\xi)\,(T\,(1-\eta) - 2\,P_E\,\eta)\right) \quad \text{(H.9)}$$

$$\tau_T^+ = \frac{U^2}{2}\frac{1}{3\,P_F^2\,\rho_E^2}\times \\ \left(P_E\,\rho_E\,(4-\eta)\,(2\,T(1-\xi) - P_F\,\xi) - P_E^2\,(1-\eta)(3-\xi)\right. \\ \left. + 6\,\rho_E^2\,(P_F-T)\,(P_F + T - T\,\xi)\right) \quad \text{(H.10)}$$

where Θ is the Heaviside function. As already discussed in the main text, τ_T^- is zero for $T > P_F/\rho_F$. Similar to the previous results all values are independent of S as long as $S > P_E$. In order to calculate the γ_T value, we need to include κ_T and $\tau_T^{\pm}$ into equation 4.26.

H.7 ANALYTICAL CALCULATION OF γ_T FOR THE S&B MODEL

Additionally, we find that if we set $\rho_E = \rho_F$ and use equation H.1 for the signal shape, κ_T is always greater than zero for $\rho_E > 1$ and thus, γ_T never diverges. This becomes clear if we investigate κ_T for $\rho_E = \rho_F$:

$$\kappa_T = \frac{U^2}{2} \frac{\rho_E - 1}{\rho_E^2} \left(6\left(1 + \rho_E\right) - \left(4 - \eta\right)\eta - \left(4 - \xi\right)\xi \right) \tag{H.11}$$

As ρ_E is greater than 1, $\rho_E - 1$ is always greater than zero and the first two products are positive values. Therefore we need to make sure that the last product is positive, too. The first summand is always greater than 12 ($\rho_E > 1$) and, as η and ξ is bounded between 0 and 2, the second and third summand are each smaller 4. This shows that the last product is always larger than $12 - 4 - 4 = 4$ and γ_T is always convergent.

H.7 Analytical calculation of γ_T for the S&B model

In the S&B model the control parameters ρ_E and ρ_F are both infinite, which leads to a rising and a falling phase of length 0. This means that we have rectangular-like signal shapes. Due to this property we get for the derivative of the signal u two δ functions at $t = 0$ and $t = S$, where the second is negative. Thus, the integral yielding κ_T (equation 4.33) simplifies to $-(u(0) - u(S)) = U$, with U being the height of our general signal function u. Note that we are not restricted to signal shapes given by equation H.1. We also use the two δ functions to solve the $\tau_T^{\pm}$-integrals. This leads to $\tau_T^- = u(-T)$ and $\tau_T^+ = u(S + T)$. Taking this results, we calculate γ_T to

$$\gamma_{TSB} = \left(\frac{U}{2\,u(S+T)} + \sqrt{\frac{U^2}{4\,u^2(S+T)} + \frac{u(-T)}{u(S+T)}} \right)^{-1} \tag{H.12}$$

or, even simpler, if we restrict our system to $T > 0$, we get $\gamma_{TSB} = u(S+T)/U$. As mentioned before, this holds for all possible signal shapes on condition that the shape consists of a rising phase, a plateau and a falling phase. Additionally, as $u(t) \leq U$, γ_{TSB} is always less or equal than 1, too.

APPENDIX H ANALYTICAL CALCULATION OF γ USING FIRST AND SECOND ORDER TERMS

List of Symbols

Symbol		Range	Description/Comment
η, ξ, ν		$\mathbb{R}$	real-valued, used in various contexts
n, k		$\mathbb{N}$	number, used in various contexts
$\dot{\eta}(t)$	$\frac{d}{dt}\eta(t)$	$\mathbb{R}$	temporal derivative
$\boldsymbol{\eta}$		$\mathbb{R}$	matrix
$\delta(t)$			delta-function
F, G			functionals
$*$			convolution
General symbols			
i, j		$\mathbb{N}^+$	neuron
N		$\mathbb{N}$	number of neurons
t, z		$\mathbb{R}$	time (z used as integration variable)
T	$t_i - t_j$	$\mathbb{R}$	temporal difference between events
x_i		$\mathbb{R}$	unfiltered input
u_i	$(x_i * h)$	$\mathbb{R}$	filtered input
v		$\mathbb{R}$	output
w_i		$\mathbb{R}$	weight
$\Delta w_i, \Delta_i$		$\mathbb{R}$	weight *change* after event
$w_i^\infty, \hat{w}_i$		$\mathbb{R}$	weight after event
ac, cc		$\mathbb{R}$	auto- and cross-correlation
$\Delta w^{ac}, \Delta w^{cc}$		$\mathbb{R}$	auto- and cross-correlation contribution
μ, α		$\mathbb{R}^+, \ll 1$	plasticity/learning rate
h_i		$\mathbb{R}$	filter/kernel function
a, b, σ	$b > a$	$\mathbb{R}^+$	kernel parameters
ρ	a_v/a	$\mathbb{R}^+$	ratio indicating variability of output trace
r		$\mathbb{R}$	reward
γ		$\mathbb{R}$	discount factor

List of Symbols

Symbol		Range	Description/Comment
Chapter 2			
R		$\mathbb{R}^+$	relevance signal
T_R		$\mathbb{R}$	timing of the relevance signal
δ_r		$\mathbb{R}$	δ error
$\tilde{r}$		$\mathbb{R}$	reward amplitude
Chapter 3			
Λ	$W\,\mu\,w_0\,\dot{u}(0)$	$\mathbb{R}^+$	constant factor (W = number of events)
$\boldsymbol{A}, \mathfrak{A}$		$\mathbb{R}$	matrix describing weight change and its integral
$\boldsymbol{\Omega}$		$\mathbb{R}$	Magnus series
$\mathfrak{B}$		$\mathbb{R}$	matrix describing weight development
k		$\mathbb{N}^+$	degree of approximation
$\hat{\eta}$		$\mathbb{R}$	value after event
Chapter 4			
R		$\mathbb{R}$	Return
M		0, 1	modulatory/third factor
S		$\mathbb{R}^+$	state duration
T		$\mathbb{R}^+$	time between two consecutive states
O		$\mathbb{R}$	onset of third factor
L		$\mathbb{R}^+$	duration of third factor
P_R, P_F		$\mathbb{R}^+$	duration of rising and falling phase
P	$P_R = P_F$	$\mathbb{R}^+$	duration of both rising and falling phase
π		$\mathbb{R}^+$	bounded temporal path
κ	$-\Delta w^{ac}$	$\mathbb{R}$	negative auto-correlation contribution
τ	$\propto cc$	$\mathbb{R}$	proportional to cross-correlation
$ac+$, $ac-$		$\mathbb{R}$	auto-correlation of rising and falling phase
κ^+, κ^-	$-(\kappa^+ + \kappa^-)$	$\mathbb{R}^+, \mathbb{R}^-$	κ value of rising and falling phase
$cc-$, $cc+$		$\mathbb{R}$	cross-correlation with previous and next state
τ^-, τ^+		$\mathbb{R}^-, \mathbb{R}^+$	τ value related to previous and next state
$\gamma^\pm$	$\tau^\pm/\kappa$	$\mathbb{R}$	particular γ value, also discount factor
Chapter 5			
a_i		$\mathbb{N}$	action
d_i		$\mathbb{R}$	dendrite
D		$\mathbb{N}^+$	number of dendrites

Symbol	Range	Description/Comment
Indices		
η_v		variable output trace
η_R		relevance signal
η_G		global third factor
η_L		local third factor
η_T		different time scales
η_E		rising phase
η_F		falling phase

Bibliography

Baird, L. (1993). Advantage updating. Technical Report WL-TR-93-1146, Wright Laboratory, Wright-Patterson Air Force Base, OH 45433-7301, USA.

Balkenius, C. and Morén, J. (1998). *Computational models of classical conditioning: a comparative study*. Lecture Notes in Science. Cambridge, MA; MIT Press.

Barbour, B., Brunel, N., Hakim, V., and Nadal, J.-P. (2007). What can we learn from synaptic weight distributions? *Trends in Neurosciences*, 30 (12):622–629.

Baxter, J., Bartlett, P. L., and Weaver, L. (2001). Experiments with infinite-horizon,policy-gradient estimation. *Journal of Artificial Intelligence Research*, 15:351–381.

Bi, G. and Poo, M. (1998). Synaptic modifications in cultured hippocampal neurons: Dependence on spike timing, synaptic strength, and postsynaptic cell type. *Journal of Neuroscience*, 18(24):10464–10472.

Bi, G. Q. (2002). Spatiotemporal specificity of synaptic plasticity: cellular rules and mechanisms. *Biological Cybernetics*, 87:319–332.

Bi, G.-Q. and Poo, M. (2001). Synaptic modification by correlated activity: Hebb's postulate revisited. *Annual Review of Neuroscience*, 24:139–166.

Bienenstock, E., Cooper, L. N., and Munro, P. (1982). Theory for the development of neuron selectivity: orientation specificity and binocular interaction in visual cortex. *Journal of Neuroscience*, 2(2):23–48.

Bliss, T. V. P. and Lømo, T. (1973). Long-lasting potentiation of synaptic transmission in the dentate area of the anaesthetized rabbit following stimulation of the perforant path. *Journal of Physiology*, 232(2):331–356.

Boykina, T. B. (2003). Derivatives of the dirac delta function by explicit construction of sequences. *American Journal of Physiology*, 71 (5):462 – 468.

Braitenberg, V. (1984). Vehicles.

Burkitt, A. N., Gilson, M., and van Hemmen, J. L. (2007). Spike-timing-dependent plasticity for neurons with recurrent connections. *Biological Cybernetics*, 96:533–546.

Butz, M., Teuchert-Noodt, G., Grafen, K., and van Ooyen, A. (2008). Inverse relationship between adult hippocampal cell proliferation and synaptic rewiring in the dentate gyrus. *Hippocampus*, 18(9):879–898.

Chklovskii, D., Mel, B. W., and Svoboda, K. (2004). Cortical rewiring and information storage. *Nature*, 431:782–788.

Colbert, C. (2001). Back-propagating action potentials in pyramidal neurons: a putative signaling mechanism for the induction of hebbian synaptic plasticity. *Restorative Neurology and Neuroscience*, 19(3-4):199–211.

Dayan, P. (2002). Matters temporal. *TRENDS in Cognitive Sciences*, 6(3):105–106.

Dayan, P. and Abbott, L. F. (2001). *Theoretical Neuroscience*. Cambridge, MA; MIT Press.

Dayan, P. and Sejnowski, T. (1994). TD(λ) converges with probability 1. *Machine Learning*, 14(3):295–301.

Doya, K. (1996). Temporal difference learning in continuous time and space. In Touretzky, D. S., Mozer, M. C., and Hasselmo, M. E., editors, *Advances in Neural Information Processing Systems*, volume 8, pages 1073–1079. The MIT Press.

Doya, K. (2000). Complementary roles of basal ganglia and cerebellum in learning and motor control. *Current Opinion in Neurobiology*, 10(6).

Dudek, S. and Bear, M. (1992). Homosynaptic Long-Term Depression in Area CA1 of Hippocampus and Effects of N-Methyl-D-Aspartate Receptor Blockade. *Proceedings of the National Academy of Sciences*, 89(10):4363–4367.

Dudek, S. and Bear, M. (1993). Bidirectional long-term modification of synaptic effectiveness in the adult and immature hippocampus. *Journal of Neuroscience*, 13(7):2910–2918.

Feldman, D. E. (2000). Timing-based LTP and LTD at vertical inputs to layer II/III pyramidal cells in rat barrel cortex. *Neuron*, 27:45–56.

Fiorillo, C. D., Tobler, P. N., and Schultz, W. (2003). Discrete coding of reward probability and uncertainty by dopamine neurons. *Science*, 299(5614):1898–1902.

Florian, R. V. (2007). Reinforcement learning through modulation of spike-timing-dependent synaptic plasticity. *Neural Computation*, 19:1468–1502.

Fox, K. and Wong, R. O. (2005). Comparison of experience-dependent plasticity in the visual and somatosensory systems. *Neuron*, 48:465–477.

Gerstner, W., Kempter, R., van Hemmen, L., and Wagner, H. (1996). A neuronal learning rule for sub-millisecond temporal coding. *Nature*, 383:76–78.

BIBLIOGRAPHY

Gerstner, W. and Kistler, W. (2002a). *Spiking Neuron Models: An Introduction*. Cambridge University Press, New York, NY, USA.

Gerstner, W. and Kistler, W. M. (2002b). Mathematical formulations of Hebbian learning. *Biological Cybernetics*, 87:404–415.

Golding, N., NP., S., and N., S. (2002). Dendritic spikes as a mechanism for cooperative long-term potentiation. *Nature*, 418(6895):326–331.

Graybiel, A. (1998). The basal ganglia and chunking of action repertoires. *Neurobiology of Learning and Memory*, 70(1-2):119–36.

Hassani, O. K., Cromwell, H. C., and Schultz, W. (2001). Influence of Expectation of Different Rewards on Behavior-Related Neuronal Activity in the Striatum. *Journal of Neurophysiology*, 85(6):2477–2489.

Hebb, D. O. (1949). *The organization of behavior: A neuropsychological theory*. Wiley, Oxford, England.

Hertz, J., Krogh, A., and Palmer, R. G. (1991). *Introduction to the theory of neural computation*. Addison Wesley.

Hopfield, J. J. (1982). Neural networks and physical systems with emergent collective computational properties. *Proceedings of the National Academy of Sciences of the United States of America*, 79:2554–2558.

Hull, C. L. (1939). The problem of stimulus equivalence in behavior theory. *Psychological Review*, 46:9–30.

Hull, C. L. (1943). *Principles of Behavior*. Appleton Century Crofts, New York.

Humeau, Y., Shaban, H., Bissiere, S., and Luthi, A. (2003). Presynaptic induction of heterosynaptic associative plasticity in the mammalian brain. *Nature*, 426(6968):841–845.

Izhikevich, E. (2007). Solving the distal reward problem through linkage of STDP and dopamine signaling. *Cerebral Cortex*, 17:2443–2452.

Joel, D., Niv, Y., and Ruppin, E. (2002). Actor-critic models of the basal ganglia: new anatomical and computational perspectives. *Neural Networks*, 15:535–547.

Kayser, C., Salazar, R., and König, P. (2003). Reponses to natural scenes in cat V1. *Journal of Neurophysiology*, 90:1910–20.

Kempter, R., Gerstner, W., and van Hemmen, J. L. (2001). Intrinsic stabilization of output rates by spike-based hebbian learning. *Neural Computation*, 13(12):2709–2741.

Klopf, A. H. (1972). Brain function and adaptive systems - a heterostatic theory. Technical report, Air Force Cambridge Research Laboratories Special Report No. 133, Defense Technical Information Center, Cameron Station, Alexandria, VA 22304.

Klopf, A. H. (1982). *The hedonistic neuron: A theory of memory, learning, and intelligence.* Hemisphere, Washington, DC.

Klopf, A. H. (1988). A neuronal model of classical conditioning. *Psychobiology*, 16(2):85–123.

Kolodziejski, C., Porr, B., and Wörgötter, F. (2006). Fast, flexible and adaptive motor control achieved by pairing neuronal learning with recruitment. In *Proceedings of the fifteenth annual computational neuroscience meeting CNS*2006, Edinburgh*. Proceedings of the fifteenth annual computational neuroscience meeting CNS*2006, Edinburgh.

Kolodziejski, C., Porr, B., and Wörgötter, F. (2007). Anticipative adaptive muscle control: Forward modeling with self-induced disturbances and recruitment. In *BMC Neuroscience 2007, 8(Suppl 2)*, page 202. Proceedings of the fifteenth annual computational neuroscience meeting CNS*2007, Toronto.

Kolodziejski, C., Porr, B., and Wörgötter, F. (2008). On the equivalence between differential hebbian and temporal difference learning. In *Proceedings of the Computational and Systems Neuroscience meeting COSYNE*2008, Salt Lake City*. Proceedings of the Computational and Systems Neuroscience meeting COSYNE*2008, Salt Lake City.

Kosco, B. (1986). Differential Hebbian learning. In Denker, J. S., editor, *Neural Networks for Computing: AIP Conference Proceedings.*, volume 151. New York: American Institute of Physics.

Kulvicius, T., Porr, B., and Wörgötter, F. (2007). Chained learning architectures in a simple closed-loop behavioural context. *Biological Cybernetics*, 97(5):363–378.

Kushner, H. K. and Clark, D. S. (1978). *Stochastic Approximation for Constrained and Unconstrained Systems.* Berlin: Springer-Verlag.

Lisman, J. (1989). A mechanism for the hebb and the anti-hebb processes underlying learning and memory. *Proceedings of the National Academy of Sciences of the United States of America*, 86(23):9574–9578.

Magee, J. C. and Johnston, D. (1997). A synaptically controlled, associative signal for Hebbian plasticity in hippocampal neurons. *Science*, 275:209–213.

Magnus, W. (1954). On the exponential solution of differential equations for a linear operator. *Communications on pure and applied mathematics*, VII:649 – 673.

Malenka, R. C. and Nicoll, R. A. (1999). Long-term potentiation-a decade of progress? *Science*, 285:1870–1874.

Manoonpong, P., Geng, T., Kulvicius, T., Porr, B., and Wörgötter, F. (2007). Adaptive, Fast Walking in a Biped Robot under Neuronal Control and Learning. *PLoS Computational Biology*, 3(7):e134.

Markram, H., Lübke, J., Frotscher, M., and Sakmann, B. (1997). Regulation of synaptic efficacy by coincidence of postsynaptic APs and EPSPs. *Science*, 275:213–215.

Martin, S. J. and Morris, R. (2002). New life in an old idea: the synaptic plasticity and memory hypothesis revisited. *Hippocampus*, 12(5):609–636.

Miller, J. D., Sanghera, M. K., and German, D. C. (1981). Mesencephalic dopaminergic unit activity in the behaviorally conditioned rat. *Life Sciences*, 29:1255–1263.

Miller, K. D. and MacKay, D. J. C. (1994). The role of constraints in hebbian learning. *Neural Computation*, 6(1):100–126.

Montague, P., Dayan, P., Person, C., and Sejnowski, T. (1995). Bee foraging in uncertain environments using predictive hebbian learning. *Nature*, 376:725–728.

Montague, P. R., Dayan, P., and Sejnowski, T. J. (1996). A framework for mesencephalic dopamine systems based on predictive hebbian learning. *Journal of Neuroscience*, 76(5):1936–1947.

Morris, G., Arkadir, D., Nevet, A., Vaadia, E., and Bergman, H. (2004). Coincident but distinct messages of midbrain dopamine and striatal tonically active neurons. *Neuron*, 43(1):133–43.

Morris, G., Nevet, A., Arkadir, D., Vaadia, E., and Bergman, H. (2006). Midbrain dopamine neurons encode decisions for future action. *Nature Neuroscience*, 9 (8):1057–1063.

Morris, R. G. (1989). Synaptic plasticity and learning: selective impairment of learning rats and blockade of long-term potentiation in vivo by the n-methyl-d- aspartate receptor antagonist ap5. *Journal of Neuroscience*, 9:3040–3057.

Oja, E. (1982). A simplified neuron model as a principal component analyzer. *Journal of Mathematical Biology*, 15(3):267–273.

Pavlov, P. I. (1927). *Conditioned reflexes*. Oxford University Press, London.

Pawlak, V. and Kerr, J. N. D. (2008). Dopamine Receptor Activation Is Required for Corticostriatal Spike-Timing-Dependent Plasticity. *Journal of Neuroscience*, 28(10):2435–2446.

Porr, B., Saudargiene, A., and Wörgötter, F. (2004). Analytical solution of spike-timing dependent plasticity based on synaptic biophysics. In *Advances in Neural Information Processing Systems 17*, volume 16. MIT Press.

Porr, B., von Ferber, C., and Wörgötter, F. (2003). ISO Learning Approximates a Solution to the Inverse-Controller Problem in an Unsupervised Behavioral Paradigm. *Neural Computation*, 15:865–884.

Porr, B. and Wörgötter, F. (2003a). Isotropic Sequence Order Learning. *Neural Computation*, 15:831–864.

Porr, B. and Wörgötter, F. (2003b). Isotropic-sequence-order learning in a closed-loop behavioural system. *Philosophical Transaction of the Royal Society of London A*, 361:2225–2244.

Porr, B. and Wörgötter, F. (2006). Strongly improved stability and faster convergence of temporal sequen ce learning by utilising input correlations only. *Neural Computation*, 18:1380–1412.

Porr, B. and Wörgötter, F. (2007). Learning with "relevance": Using a third factor to stabilise hebbian learning. *Neural Computation*, 19:2694–2719.

Potjans, W., Morrison, A., and Diesmann, M. (2009). A spiking neural network model of an actor-critic learning agent. *Neural Computation*, 21:301–339.

Rall, W. (1967). Distinguishing theoretical synaptic potentials computed for different soma-dendritic distributions of synaptic input. *Journal of Neurophysiology*, 30:1138–1168.

Rao, R. and Sejnowski, T. (2001). Spike-timing-dependent hebbian plasticity as temporal difference learning. *Neural Computation*, 13:2221–2237.

Redgrave, P. and Gurney, K. (2006). The short-latency dopamine signal: a role in discovering novel actions? *Nature Reviews Neuroscience*, 7:967–975.

Rescorla, R. A. and Wagner, A. R. (1972). A theory of pavlovian conditioning: Variations on the effectiveness of reinforcement and nonreinforcement. In *In A. H. Black & W. F. Prokasy (Eds.), Classical conditioning: II. Current research and theory (pp. 64-99)*.

Rioult-Pedotti, M. S., Friedman, D., Hess, G., and Donoghue, J. P. (1998). Strengthening of horizontal cortical connections following skill learning. *Nature Neuroscience*, 1(3):230–234.

Roberts, P. (1999). Computational consequences of temporally asymmetric learning rules: I. differential hebbian learning. *Journal of Computational Neuroscience*, 7(3):235–46.

Roberts, P., Santiago, R., and Lafferriere, G. (2009). An implementation of reinforcement learning based on spike-timing dependent plasticity. *Biological Cybernetics*, 99(6):517–523.

Roberts, P. D. (2000). Dynamics of temporal learning rules. *Physical Review E*, 62(3):4077–4082.

Saudargiene, A., Porr, B., and Wörgötter, F. (2004). How the shape of pre- and postsynaptic signals can influence STDP: a biophysical model. *Neural Computation*, 16:595–626.

Saudargiene, A., Porr, B., and Wörgötter, F. (2005). Local learning rules: predicted influence of dendritic location on synaptic modification in spike-timing-dependent plasticity. *Biological Cybernetics*, 92:128–138.

BIBLIOGRAPHY

Schultz, W. (1998). Predictive reward signal of dopamine neurons. *Journal of Neurophysiology*, 80:1–27.

Schultz, W., Apicella, P., Scarnati, E., and Ljungberg, T. (1992). Neuronal activity in monkey ventral striatum related to the expectation of reward. *Journal of Neuroscience*, 12(12):4595–610.

Schultz, W., Dayan, P., and Montague, P. R. (1997). A neural substrate of prediction and reward. *Science*, 275:1593–1599.

Singh, S. P., Jaakkola, T., Littman, M. L., and Szepesvári, C. (2000). Convergence results for single-step on-policy reinforcement-learning algorithms. *Machine Learning*, 38(3):287–308.

Singh, S. P. and Sutton, R. S. (1996). Reinforcement learning with replacing eligibility traces. *Machine Learning*, 22:123–158.

Skinner, B. F. (1933). The rate of establishment of a discrimination. *Journal of General Psychology*, 9:302–350.

Spudich, J. L. and Koshland, D. E. J. (1975). Quantitation of the sensory response in bacterial chemotaxis. *Proceedings of the National Academy of Sciences of the United States of America*, 72:710–713.

Suri, R. E., Bargas, J., and Arbib, M. A. (2001). Modeling functions of striatal dopamine modulation in learning and planning. *Neuroscience*, 103(1):65–85.

Suri, R. E. and Schultz, W. (1998). Learning of sequential movements by neural network model with dopamine-like reinforcement signal. *Experimental Brain Research*, 121:350–354.

Suri, R. E. and Schultz, W. (1999). A neural network model with dopamine-like reinforcement signal that learns a spatial delayed response task. *Journal of Neuroscience*, 91(3):871–890.

Suri, R. E. and Schultz, W. (2001). Temporal difference model reproduces anticipatory neural activity. *Neural Computation*, 13(4):841–62.

Sutton, R. and Barto, A. (1981). Towards a modern theory of adaptive networks: Expectation and prediction. *Psychological Review*, 88:135–170.

Sutton, R. and Barto, A. (1998). *Reinforcement Learning: An Introduction*. MIT Press, Cambridge, MA.

Sutton, R. S. (1988). Learning to predict by the method of temporal differences. *Machine Learning*, 3:9–44.

Sutton, R. S. and Barto, A. G. (1990). Time-derivative models of pavlovian reinforcement. In *Learning and Computational Neuroscience: Foundations of Adaptive Networks*, pages 497–537. MIT Press.

Tamosiunaite, M., Ainge, J., Kulvicius, T., Porr, B., Dudchenko, P., and Wörgötter, F. (2008). Path-finding in real and simulated rats: On the usefulness of forgetting and frustration for navigation learning. *Journal of Computational Neuroscience*, 25(3):562–582.

Tamosiunaite, M., Porr, B., and Wörgötter, F. (2007). Developing velocity sensitivity in a model neuron by local synaptic plasticity. *Biological Cybernetics*, 96:507–518.

Thompson, A. M., Porr, B., Kolodziejski, C., and Wörgötter, F. (2008). Second order conditioning in the sub-cortical nuclei of the limbic system. In *SAB '08: Proceedings of the 10th international conference on Simulation of Adaptive Behavior*, pages 189–198. Springer-Verlag, Berlin, Heidelberg.

Thorndike, E. L. (1933). A theory of the action of the after-effects of a connection upon it. *Psychological Review*, 40:434–439.

Tobler, P. N., Fiorillo, C. D., and Schultz, W. (2005). Adaptive coding of reward value by dopamine neurons. *Science*, 307(5715):1642–1645.

Tsitsiklis, J. N. and Van Roy, B. (1997). An Analysis of Temporal-Difference Learning with Function Approximation. *IEEE Transactions on Automatic Control*, 42(5):674–690.

Tsukamoto, M., Yasui, T., Yamada, M. K., Nishiyama, N., Matsuki, N., and Ikegaya, Y. (2003). Mossy fibre synaptic NMDA receptors trigger non-Hebbian long-term potentiation at entorhino-CA3 synapses in the rat. *Journal of Physiology*, 546(3):665–675.

van Rossum, M. C. W., Bi, G. Q., and Turrigiano, G. G. (2000). Stable hebbian learning from spike timing-dependent plasticity. *Journal of Neuroscience*, 20(23):8812–8821.

Watkins, C. and Dayan, P. (1992). Technical note:Q-Learning. *Machine Learning*, 8:279–292.

Wiering, M. (2004). Convergence and divergence in standard averaging reinforcement learning. In Boulicaut, J., Esposito, F., Giannotti, F., and Pedreschi, D., editors, *Proceedings of the 15th European Conference on Machine learning ECML'04*, pages 477–488.

Witten, I. H. (1977). An adaptive optimal controller for discrete-time markov environments. *Information and Control*, 34:286–295.

Yang, S. N., Tang, Y. G., and Zucker, R. S. (1999). Selective induction of LTP and LTD by postsynaptic Ca^{2+} elevation. *Journal of Neurophysiology*, 81:781–787.

Acknowledgments

This thesis would not have been possible without the support of many friends and colleagues. Hence, I would like to thank Prof. Dr. Florentin Wörgötter for his guide and advice. Many fruitful discussions with him have always led onwards. I also thank him for the opportunity to present my research at conferences and for the possibility of research visits (Karlsruhe, Glasgow, London). Next, I would like to thank Prof. Dr. Theo Geisel for the stimulating and encouraging working condition at the Max-Planck-Institute. Special thanks goes to Dr. Bernd Porr who is coauthor of almost all of my publications. The discussions I had with him were always inspiring and many ideas leading to this thesis were developed during my visits to Glasgow. Also Prof. Dr. Minija Tamosiunaite receives many thanks for her help during several stages of this work. Further on, I thank the Bernstein Center for Computational Neuroscience for financing this work.

Another important part of the acknowledgments is dedicated to all members of Florentin's group which are Babette Dellen, Bettina Hoffmann, Irene Markelić, Nataliya Shylo, Kristin Stamm, Silke Steingrube Ursula Hahn-Wörgötter (who was always a great help with administration), Alexey Abramov, Eren Erdal Aksoy, Jan-Matthias Braun, Markus Butz, Sinan Kalkan, Tomas Kulvicius, Guoliang Liu, Poramate Manoonpong, Kejun Ning, Johannes Schröder-Scheteling, Daniel Steingrube, Harm-Friedrich Steinmetz, Christian Tetzlaff, Steffen Wischmann, and Alexander Wolf. It was a great pleasure to be part of this group. It was more than a research group, which can be seen from the ski trips and the daily after-lunch kicker matches. Along the same line I also would like to thank the members of the Nonlinear Dynamics group, in particular Katharina Jeremias, Katja Fiedler, Tanja Gindele, Anna Levina, Corinna Trautsch, Regina Wunderlich, Yorck-Fabian Beensen, Vitaly Belik, Armin Bies, Denny Fliegner, Michael Herrmann, Frank Hesse, Georg Martius, Raoul Martin Memmesheimer, Tobias Niemann, Michael Schnabel, Hecke Schrobsdorf, Marc Timme, and Fred Wolf.

This thesis benefited substantially from the proofreaders and therefore I thank Kelly Paschal, Alexey Abramov, Eren Erdal Aksoy Tomas Kulvicius, and Harm-Friedrich Steinmetz very much for their effort.

During my visits to Glasgow, as already mentioned above, many ideas arouse that are now part of this thesis. For this great time I thank Adedoyin Maria Thompson, Lynsey McCabe and Paolo Di Prodi.

And, not to be forgotten, I would like to thank all the friends from the years of my study in Würzburg which are Anna Tschetschetkin, Christian Schmidt, Irene und Marcel Schumm, Martina und Marc Wisniewski, Christian Weigand, Andy Bolzmann und Eva Wiese for their friendship and their ongoing support, and also my former girlfriend Franziska Klingner for listening to all the problems which came up while working on this thesis and her emotional support.

And last but definitely not least, I would like to thank my whole family, in particular my mother Lydia, who supported me with all her energy on the way to as well as during my graduation. A special thanks goes to my grandmother Gertrud who can not be with us to see the publishing of this thesis.

Die VDM Verlagsservicegesellschaft sucht für wissenschaftliche Verlage abgeschlossene und herausragende

Dissertationen, Habilitationen, Diplomarbeiten, Master Theses, Magisterarbeiten usw.

für die kostenlose Publikation als Fachbuch.

Sie verfügen über eine Arbeit, die hohen inhaltlichen und formalen Ansprüchen genügt, und haben Interesse an einer honorarvergüteten Publikation?

Dann senden Sie bitte erste Informationen über sich und Ihre Arbeit per Email an *info@vdm-vsg.de*.

Sie erhalten kurzfristig unser Feedback!

VDM Verlagsservicegesellschaft mbH
Dudweiler Landstr. 99
D - 66123 Saarbrücken
Telefon +49 681 3720 174
Fax   +49 681 3720 1749
www.vdm-vsg.de

Die VDM Verlagsservicegesellschaft mbH vertritt

Printed by Books on Demand GmbH, Norderstedt / Germany